高等教育（矿业）"十三五"规划教材
中国矿业大学教学名师培育工程资助项目
大学英语ESP系列教材

矿业工程概况

A Guide to Mining Engineering

主　编　朱　哲　沈　丛
副主编　赵　君　何　暄
编　者　黄　敏　杜光明　温力亚
　　　　赵　虹　徐文东

中国矿业大学出版社
China University of Mining and Technology Press

图书在版编目(CIP)数据

矿业工程概况 = A Guide to Mining Engineering：英文 / 朱哲，沈丛主编. —徐州：中国矿业大学出版社，2017.6
ISBN 978-7-5646-3528-2

Ⅰ.①矿… Ⅱ.①朱… ②沈… Ⅲ.①矿业工程—英语—高等学校—教材 Ⅳ.①TD

中国版本图书馆 CIP 数据核字(2017)第 103823 号

书　　名	矿业工程概况
主　　编	朱哲 沈丛
责任编辑	万士才
出版发行	中国矿业大学出版社有限责任公司
	（江苏省徐州市解放南路　邮编 221008）
营销热线	(0516)83885307　83884995
出版服务	(0516)83885767　83884920
网　　址	http：//www.cumtp.com　E-mail：cumtpvip@cumtp.com
印　　刷	徐州中矿大印发科技有限公司
开　　本	787×960　1/16　印张 10.25　字数 195 千字
版次印次	2017 年 6 月第 1 版　2017 年 6 月第 1 次印刷
定　　价	16.50 元

（图书出现印装质量问题，本社负责调换）

前 言

2007年教育部正式发布了《大学英语课程教学要求》,对大学英语教学的目标进行了清楚的界定:大学英语的教学目标是培养学生的英语综合应用能力,特别是听说能力,使他们在今后学习、工作和社会交往中能用英语有效地进行交际,同时增强其自主学习能力,提高综合文化素养,以适应我国社会发展和国际交流的需要。《大学英语课程教学要求》对英语综合应用能力进行了清楚地阐释,即学生在学习、工作和社会交往中用英语进行交际的能力,由此可见大学公共英语教学的方向最终还是朝着实用性迈进。

如何结合中国矿业大学的优势学科,为社会培养具备扎实外语功底的复合型人才,不仅是学生学习的需求,也是高校在竞争中获胜的需求。为此,我们组织教师编写了大学英语ESP系列教材之《矿业工程概况》,以期为推动大学英语课程教学改革进行实践探索。本教材的指导思想和特点主要体现在以下几点:

首先,在探索语言知识与专业学科结合的语言教学过程中,强调内容与语言整合型的语言教学方法,希望将目前通用英语导向的大学英语教学转变为培养学生用英语从事专业学习或是应对未来职业的专门用途英语,拓展学生的国际视野,提升专业领域内的跨文化交流能力、学术沟通和合作能力以及参与国际竞争的能力。

其次,突出行业特色和优势学科。结合我校行业特色和优势学科,本教材选取"矿业工程"作为教材编写主题,采取英语与行业特色、优势学科和专业相结合的方式,使学生能在有限时间内最大限度地获取胜任将来工作需要和综合素质提高的专业相关知识。

第三,本教材以实用为导向,与学生专业紧密结合,提出整个英语教学要以学生需要,尤其是专业需求为中心,认为要把英语作为手段或工具来学习运用,而不仅仅作为一门语言课程。因此在设计ESP系列教材时,我们并未强调由低到高的渐进式学习模式,而是首先分析不同学习者的需要,充分考虑与目标达成度相适应的不同能力水平,并以此为依据,结合学习者的基础,确立相应的教学目标、选材内容以及测试方法等,以培养学生在有限时间内最大限度获取胜任将来工作所需的英语能力。

第四,以学习者为中心的编写理念。本教材编写基于高校国际化办学目标、助力优势学科发展、满足学生需求的"以学生为中心"的理念之上,强调把学习和技能的提高放在首位,充分考虑高校的行业和专业特色、学生本身的英语能力及其差异化的语言学习需求。

第五,本教材主题内容的确定充分吸纳了我校矿业工程专业老师的建议,涵盖了矿业工程的基本概念和基础知识,内容专业性较强,权威性较高。本教材中,专业术语的出现是建立在高频统计基础之上的,强调以实证为依据,而非以往以直觉为基础来安排教学内容。

本教材的使用对象是高等院校的非英语专业学生,共8个单元,每个单元包括3个部分,第一部分为主课文及练习。主课文内容主要涵盖我国煤炭矿业发展现状及面临的挑战,习题设计紧扣课文内容,让学生能够学练结合,举一反三,达到语言学习、专业知识学习和文化学习的目的;第二部分为副课文,选取的素材主要围绕我国和世界煤炭开采的基本情况,难度介于专业和普通英语之间,习题的设计主要以理解关键信息为主;第三部分为知识拓展和延伸,提供了一篇篇幅较长的文章供学生阅读,以期了解煤炭矿业的发展前景,并配有拓展词汇供学习者参考。

本教材受中国矿业大学教学名师培育工程资助,并得到了2013年江苏省高等教育教改课题"基于煤炭矿业特色的大学英语ESP系列教材建设研究"部分资助,由中国矿业大学外文学院朱哲和沈丛负责组织编写,具体编写分工如下:第1单元由黄敏和徐文东负责编写,第2单元由赵君负责编写,第3、4单元由何暄负责编写,第5、6单元由温力亚负责编写,第7单元由朱哲和沈丛负责编写,第8单元由杜光明负责编写。另外,赵虹教授在材料审定和全书审校方面提供了诸多帮助,在此一并表示感谢。

鉴于编者水平有限,错误和不足之处在所难免,欢迎教材使用者提出批评、意见和建议。

朱 哲

2017年5月

目 录

Unit 1
    Formation of Coal ··· 1

Unit 2
    A Kaleidoscope of Coal Mines ································· 18

Unit 3
    Coal Industry and Chinese Social Development ············ 38

Unit 4
    Mining Culture ·· 60

Unit 5
    Chinese Coal Mining Industry ································· 78

UINT 6
    Five Major Mining Disasters ·································· 93

Unit 7
    Ecological Crisis ·· 109

Unit 8
    Clean Coal Technology ·· 137

Unit 1　Formation of Coal

Foreword

In this unit, you will learn about the process of coal formation. As we all know, coal has meant "mineral of fossilized carbon" since the 13th century. It is a combustible black or brownish-black sedimentary rock usually occurring in rock strata in layers or veins called coal beds or coal seams. The harder forms, such as anthracite coal, can be regarded as metamorphic rock because of later exposure to elevated temperature and pressure. Coal is composed primarily of carbon along with variable quantities of other elements, chiefly hydrogen, sulfur, oxygen, and nitrogen.

Section One

Preview

Text A tells us that the formation of coal undergoes quite a long time. The text begins with the development of plant life, which unfolds in front of us an amazing picture of a tree turning into coal. Through the story, we know what conditions are necessary for the formation of coal, and how plants become coal through different geological evolution.

Text A

How Coal Was Formed?

　　Many of the problems facing the miner and the mining engineer arise from the way coal was formed. It is not uncommon to find **imprint** of a **ferny** leaf or the stem of a plant in a large **lump** of coal. Such sights are a commonplace to the miner working underground with coal in the mass. They provided the first clues as to how this hard, black substance came to be buried deep under the surface of the earth.

  Plant life developed on the land surfaces of the world before animal life. First **mosses** and then ferns began to cover the **barren** surface. Later came forests of trees. The vegetation flourished, died and was renewed. Great icecaps advanced towards the **equator** and retreated in a rhythm measured by millions of years. As the outer crust of the earth cooled, vast earthquakes threw up mountain ranges or submerged land under the sea. **Landslips** cut off areas of the sea to form lakes or turned existing lakes into oceans. Amid this world-wide **turmoil**, coal was being formed.

  The wood in trees is largely made up of a compound, called **cellulose**, which is manufactured by the tree as it grows. The action of sunlight on the substance which gives leaves their green color (**chlorophyll**) enables the plant to absorb water and a gas called carbon dioxide from the air, and to combine them together. This process is so complicated that our finest scientists cannot yet copy it in the best of their laboratories, nor even fully understand how it works.

  All matter throughout the whole of our universe is made up of one or more of about ninety elements. These elements are all different from each other. The smallest particle of each element is called an atom. These are so small that there are more than a million, million, million atoms of iron in the head of an ordinary pin. Atoms of different elements can combine to form compounds. The smallest particle of a compound, which always contains atoms of different elements, is called a molecule. The carbon dioxide and water absorbed by plants from the air are both simple compounds. The molecule of water contains two atoms of hydrogen and one of oxygen. We cannot get a smaller particle of water than this molecule. If we attempted to do so, by splitting the molecule, we should have hydrogen and oxygen, not water. The molecule of carbon dioxide contains one atom of carbon and two of oxygen. Plants are able to combine these two simple molecules to form cellulose molecules which have 12,000 carbon atoms, 20,000 hydrogen atoms and 10,000 oxygen atoms!

  Dead vegetation left exposed to the air will slowly rot. In the end, the action of wind, rain and sunshine will turn it into a dusty powder. Yet it was from the woody part of dead vegetation that our coal was formed.

Unit 1 Formation of Coal

In the vast **upheavals** of millions of years ago some of the dead and decaying forests were trapped underground and held there under enormous pressure from the rocks above. Buried with this vegetation were minute living organisms called bacteria. There are many different kinds of bacteria. These particular ones do not live by breathing oxygen from the air as we do, but can absorb it from compounds like cellulose.

The large and complicated molecules forming the cellulose were broken up under this double attack from the bacteria and the heavy pressure underground. New molecules were formed and these in turn were robbed of more oxygen by the bacteria.

While this change was going on underground new forests were growing on the surface of the earth and some of these decayed and were **engulfed** in the same way. Sometimes the new forests had been growing for a comparatively short time when they were destroyed. The layer of vegetation buried under the rocks would then be shallow and, when compressed underground, would form a very thin layer of coal. Elsewhere, the forests would flourish for many thousands of years before some new disaster buried them. The resulting coal is sometimes more than a thousand feet thick.

The crust of the earth had not yet settled down when the original beds of coal were formed. New shifts and stresses in the rocks cracked some layers of coal and threw them up towards the surface. Others were buried more deeply under deposits of volcanic lava which afterwards cooled to form another layer of rock. Thus coal has been found at all levels. Some has been thrown above ground as "**outcrops**", while some is buried many thousands of feet underground.

The coal found on the surface was used in ancient times as a stone to be carved into beads and ornaments. There are signs that the Romans used it occasionally as a fuel but little interest was shown in the idea. Outcrop coal was comparatively rare and there was abundant fuel to be found from forest trees.

Vocabulary

combustible / kəm'bʌstəbl / ***adj.*** capable of igniting and burning 易燃的,可燃的

sedimentary / sedɪ'mentərɪ' / ***adj.*** resembling or containing or formed by the accumulation of sediment 沉淀的,沉积的

anthracite / 'ænθrəsaɪt / ***n.*** a hard natural coal that burns slowly and gives intense heat and very little smoke 无烟煤;硬煤

imprint / ɪm'prɪnt / ***n.*** a concavity in a surface produced by pressing 盖印;痕迹 a distinctive influence 特征 an identification of a publisher; a publisher's name along with the date and address and edition that is printed at the bottom of the title page 版权标记 ***v.*** mark or stamp with or as if with pressure 盖(印);刻上(记号) establish or impress firmly in the mind 使铭记

ferny / 'fɜːnɪ / ***adj.*** resembling ferns especially in leaf shape 蕨类的

lump / lʌmp / ***n.*** a large piece of something without definite shape 块,团 abnormal protuberance or localized enlargement 肿块 an awkward stupid person 笨拙的人 ***v.*** put together indiscriminately 结成块;成团;笨重地行走 ***v.*** group or chunk together in a certain order or place side by side 使成团,使成块;使团结在一起;把……混在一起

moss / mɔs / ***n.*** tiny leafy-stemmed flowerless plants 苔藓

barren / 'bærən / ***adj.*** not bearing offspring 不孕的; incapable of maintaining life 贫脊的,荒芜的

equator / ɪ'kweɪtə(r) / ***n.*** an imaginary line around the Earth forming the great circle that is equidistant from the northern and southern POLE 赤道 a circle dividing a sphere or other surface into two usually equal and symmetrical parts(平分球形物体表面的)圆,(任何)大圆

landslip. / 'lændslɪp / ***n.*** a slide of a large mass of dirt and rock down a mountain or cliff 山崩,塌方

turmoil / 'tɜːmɔɪl / ***n.*** a violent disturbance 动乱,混乱

cellulose / 'seljʊləʊz / ***n.*** a polysaccharide(多糖) that is the chief constituent of all plant tissues and fibers 细胞膜质,纤维素;(用于制作涂料、漆等的)纤维素化合物

chlorophyll / 'klɔːrəfɪl / ***n.*** any of a group of green pigments found in photosynthetic organisms 叶绿素

Unit 1 Formation of Coal

upheaval / ʌpˈhiːv(ə)l / **n.** a state of violent disturbance and disorder (as in politics or social conditions generally) 激变；剧变；动乱 (geology) a rise of land to a higher elevation (as in the process of mountain building) 隆起；举起，抬起

engulf / ɪnˈgʌlf / **v.** engross (oneself) fully 吸引，使全神贯注 flow over or cover completely 吞没，淹没，沉没

outcrop / ˈaʊtkrɒp / **n.** the part of a rock formation that appears above the surface of the surrounding land 露出地面的岩层 **v.** appear on the surface, come to the surface on the ground 露出

Phrases and Expressions

coal seam：煤层
metamorphic rock：变质岩
arise from：由……引起，起因于
a large lump of：一大块……
in the mass：总体上，总的来说
cut off：切断；中断
break up：打碎，破碎；结束；解散；衰落
be robbed of：被打劫；被剥夺；失去
settle down：定居；安定下来；专心于
throw up：抛起；举起；把……迅速往上推；呕吐；吐出；产生

Exercises

Ⅰ. Comprehension of the Text

Directions: Please answer the following questions according to the text.

1. What do many of the problems facing the miner and the mining engineer arise from?
2. What does the author tell us in paragraph 2?
3. Which process is too complicated for scientists to copy in the best of their laboratories, nor even fully understand how it works?
4. What is the smallest particle of each element?
5. What is the smallest particle of a compound?
6. What does the molecule of carbon dioxide contain?
7. Why has coal been found at all levels?

8. What was the function of the coal found on the surface in ancient times and who used it occasionally as a fuel?

II. Group Discussion

Directions: Discuss the following questions with your partners, using as much text information as possible in your discussion.

1. What's the significance of coal in social development and its development prospect?
2. What's the main impact of coal mining on the environment? Can you put up some useful suggestions and measures to minimize the impact?
3. Which kind of energy resources do you think will be the alternative ones?

III. Word Bank

Directions: Fill the blanks with the words on the right side of the text. For each word, you can use only once.

Coal was formed from plants and imprint of a 1._____ leaf in a large sum of coal that can be found by miners. The earth was primarily covered with 2._____ and then ferns. Later came forests. Dead 3._____ exposed to the air rotted and turned into 4._____ which formed coal. In the vast 5._____ of the earth, some decaying forests were trapped underground and buried with 6._____ which absorb oxygen from compounds like 7._____. With new forests growing on the earth's surface, some of these decayed and were 8._____ in the same way. The movement of the earth 9._____ cracked some layers of coal and threw them up towards the surface as 10. "_____"; others were buried underground.

a) bacteria
b) crust
c) upheavals
d) outcrops
e) ferny
f) cellulose
g) vegetation
h) dusty powder
i) engulfed
j) mosses
k) lump
l) mass
m) vegetarian
n) element
o) imprint

IV. Translation Practice

Directions: Translate the following sentences from English to Chinese.

1. Almost 40% of the children had been exposed to second-hand tobacco smoke at some point in their life.
2. He was knocked to the ground and robbed of his wallet.

3. One day, I'll want to settle down and have a family.
4. As to the matter you brought up, I think it should be settled as soon as possible.
5. Accidents often arise from carelessness.
6. The waves impinged on the rocks, throwing up pearl-like drops of spray.

Ⅴ. **Writing Practice**

Directions: Write a 3-paragraph passage of about 120 words with the title "A Long Journey to Coal" based on the following outline.

Outline:
1. It takes a plant millions of years to turn into coal.
2. During this long journey, what conditions are indispensable for the formation of coal?
3. Do you think the journey ends where human beings begin to excavate and use coal?

Section Two

Preview

What coal operation is adopted depends on where coal is found. Till now, many coal mining techniques are worked out in coal operation. Text B introduces three coal mining methods which are existent and popular in nowadays' coal mining field. Maybe in the future, new coal mining methods would be adopted when new coal fields are found in places different from where we've already known.

Text B

The Coal Mining Methods

Surface Mining

Coal found close to the surface can be uncovered and removed by large machines in a process that is called surface mining. Surface mining techniques account for 60 percent of coal produced in the United States — 75 percent in Western states, where some deposits are up to 100 feet thick.

Only recently has surface mining played an important role in the U. S. coal industry. The development and use of large power equipment provided the **impetus** that moved surface mining into prominence, and during the 1970s it became the leading method of coal mining.

Today's surface mines are large, intensively engineered, and highly efficient mechanized operations. When an area is to be mined, **topsoil** and **subsoil** are removed first and set aside to be used later in reclaiming the land. Then specially designed machines — **draglines**, **wheel excavators**, or large **shovels** — remove the rock and other material, called **overburden**, to expose the bed of coal. Smaller shovels load the coal into large trucks that remove the coal from the pit.

Once the coal is removed, the area is reclaimed. First the overburden and then the soils are replaced and the area is restored as nearly as possible to its original **contour**. Vegetation currently suitable to the area is planted to anchor the soil and return the land to a natural, productive state. Reclaimed lands are a valuable resource that can support farm crops, provide new wildlife habitats, enhance recreational opportunities, and even serve as sites for commercial development.

The complete mining operation is scheduled so that as one area is being mined, another is being reclaimed where the coal was removed. Thus, even at the largest surface mines only a relatively small area is disturbed by active mining at any one time. Since 1977 an estimated 2 million acres of coal mine lands have been reclaimed in this manner.

Underground Mining

Underground mining methods are used where the coal seam is too deep or the land too hilly for surface mining. Most underground mining takes place east of the Mississippi, especially in the Appalachian mountain states. Coal production was once dominated by underground mining methods, but the growth of coal mining in the West changed that. Now, only 40 percent of coal produced in the U. S. comes from underground mines.

Underground mines differ according to how the coal seam is situated with respect to the surface. If the coal deposit outcrops (appears at the surface) on a hillside, a drift mine can be driven horizontally into the seam. Where the bed of coal is relatively close to the surface, yet too deep to be recovered by surface mining, a slope mine can be constructed, with the mine shaft **slanting** down from the surface to the coal seam. The most common type is the **shaft** mine. To reach the coal, which may be as deep as 2,000 feet, vertical shafts are cut through the overburden to the coal bed, which is excavated by machines.

In deep mines, the seam is mined in carefully engineered patterns that keep as much as half of the coal in place to help support the roof of the active mining area. This "room and pillar" method requires that large columns of coal remain between mined-out areas, or rooms, which are created when the coal is mined, either by continuous mining machines or conventional methods.

The largest amount of coal taken from underground mines is produced using continuous miners. This machine has a large, rotating, drum-shaped cutting head **studded** with **carbide**-tipped teeth that break up the seam of coal. Large gathering arms on the machines **scoop** the coal directly onto a built-in **conveyor** for loading into waiting shuttle cars.

In conventional mining, a machine resembling an oversized chain saw cuts into the coal. This gives the coal an area to expand into during blasting. Holes are drilled for explosives, which blast loose large chunks of coal. Machines called loaders scoop the coal onto conveyors which dump it into shuttle cars that haul the coal out through the shaft. This traditional method of mining accounts for about 11 percent of total production.

In both continuous and conventional mining, the roof over the mined-out area is supported for safety. The most important development in roof support — both in terms of safety and cost — has been the "roof bolt". Roof bolts are long rods driven into the roof to bind several layers of weak strata into a layer strong enough to support its own weight. Roof bolts also can anchor a weak immediate roof to a strong, firm structure above. Machines are used to drill holes, position the bolts and tighten them.

Longwall Mining

An increasingly popular and more efficient means of underground mining—introduced from Europe in the early 1950s—is longwall mining technology. Longwall today accounts for about one-third of total underground coal tonnage in the U.S. In a continuous, smooth motion, a rotating shear on the mining machines moves back and forth along the face—or wall—of a block of coal, cutting the coal, which drops onto a **conveyor** and is removed from the mine. The block of coal being mined is several hundred feet wide, thus the name longwall.

  Where longwall mining machines are used, room-and-pillar arrangements are not created throughout the entire mine (although pillars of coal are left to support the roof in haulage ways used by people and machines moving about the mine). The longwall miner itself has a **hydraulically** operated steel canopy which holds up the roof and protects miners working at the face. As the miner cuts progressively deeper into the block of coal, the shield advances with it, allowing the unsupported roof in the mined-out area behind it to collapse in a controlled and safe manner.

Vocabulary

impetus / ˈɪmpɪtəs / *n.* the force or energy with which a body moves 动力;促进;势头;声势

topsoil / ˈtɒpsɔɪl / *n.* the top layer of soil 表土(层),耕作(层);耕层

subsoil / ˈsʌbsɔɪl / *n.* the soil lying immediately under the surface soil(土壤的)底土,心土

dragline / ˈdræglaɪn / *n.* a large excavator with a bucket pulled in by a wire cable 牵引绳索,索斗铲;拉索,拉铲挖掘机

wheel excavator / (h)wiːlˈekskəˌveɪtə / *n.* a large machine for removing soil from the ground, especially on a building site 轮式(斗轮式)挖土机

shovel / ˈʃʌvl / *n.* a long-handled tool for lifting and moving loose material; a part like this on a digging or earth-moving machine 铲子;挖土机或推土机的铲形部分

overburden / əʊvəˈbɜːdn / *v.* load (someone) with too many things to carry 装载过多,负担过多,使过劳; *n.* the surface soil that must be moved away to get at coal seams and mineral 表土,履盖层

contour / ˈkɒntʊə(r) / *n.* an outline, especially one representing or bounding the shape or form of something 外形,轮廓;(地图上表示相同海拔各点的)等高线;概要;电路

slant / slɑːnt / *v.* slope or lean in a particular direction;diverge from a vertical or horizontal line (使)倾斜,歪斜

shaft / ʃɑːft / *n.* a vertical passage into a mine 竖井

Unit 1 Formation of Coal

stud / stʌd / **v.** provide with or construct with studs 用螺栓支撑
carbide / ˈkɑːbaɪd / **n.** a binary compound of carbon with a more electropositive element 碳化物,硬质合金
scoop / skuːp / **vt.** take out or up with or as if with a scoop 掘,挖,铲
conveyor / kənˈveɪə / **n.** a moving belt that transports objects 传送带,输送机
hydraulically / haɪˈdrɔːlɪkəlɪ / **adv.** (by a liquid) moving in a confined space under pressure 液压地

Phrases and Expressions

play an important role in：在……中起重要作用
set aside：把……放在一旁,不理会
with respect to：关于,至于,谈到
drift mine：砂矿
slope mine：斜井
shuttle car：穿梭式机动矿车,梭车
chain saw：链锯,小型机器锯
account for：说明,解释
hold up：举起,抬起,支撑

Exercises

Ⅰ. Comprehension of the Text

Directions: Choose the right answer to each question according to the text.

1. Which statement is true about surface mining?
    A. Surface mining techniques account for 75% of coal produced in America.
    B. Surface mining technique plays an important role in coal producing in America today.
    C. During the 1970s, surface mining technique began to develop.
    D. Surface mining results in no environmental problems.
2. Which are commonly used machines in surface mining mentioned in the text?
    A. draglines, wheel excavators
    B. large shovels, cranes

C. wheel excavators, tractors

D. draglines, trucks

3. Where does most underground mining take place in America?

   A. In the northern part of the United States.

   B. In the southern part of the Appalachian mountain states.

   C. In the east of the Mississippi, especially in the Appalachian mountain states.

   D. Along the coastline in western part of the United States.

4. Why are continuous miners important in producing underground mines?

   A. Because they play important roles in environmental protection.

   B. Because they can be used in order to save energy.

   C. Because they have higher working efficiency.

   D. Because it is carefully designed to adjust to the mining environment.

5. Which statement is true about longwall mining?

   A. Longwall mining has become a major method in American mining industry.

   B. A rotating shear moves smoothly and continuously in longwall mining.

   C. Longwall mining accounts for about 50% of total underground coal tonnage.

   D. Longwall mining has great advantages in saving space and mining efficiency.

Section Three

Extended Reading

What Is Coal's Future?

With a 250- to 300-year supply of coal under our feet, the picture of coal's role in the future is bright. However, coal has a reputation to overcome: the idea that it is a dirty fuel. Modern coal combustion facilities, such as those found at many of the nation's electric power plants, use equipment to remove most of the polluting elements from coal smoke. In fact, so much is removed that one can hardly see any smoke at all coming from these "smokestacks."

Most of what shows is steam.

The dark, sooty material called fly ash that once went up the stack is now removed by filters or by devices called precipitators. With precipitators, the flue gas is passed through an electric field. The ash particles become negatively charged and are attracted to positively charged collecting plates and later removed for disposal. This method eliminates 99.5 percent of the offending material.

Coal contains sulfur, which combines with oxygen when the coal is burned to produce sulfur oxides. The effect of sulfur oxides on the environment has been the topic of significant debate for a number of years. Beginning in the 1970s, coal producers and major coal consumers initiated a number of efforts to reduce the amount of sulfur compounds emitted into the environment.

One was the use of "flue gas scrubbers," which can remove up to 95 percent of the sulfur oxides from the stream of gases produced by coal combustion before they go up the smokestack. In one process, sulfur dioxide in the flue gas reacts with a lime or limestone water slurry to form calcium sulfite or calcium sulfate (gypsum) sludge. In another, sulfur and sulfuric acid are produced as by-products. Still others can produce a dry by-product.

Sulfur emissions (SO_2 — sulfur dioxide) also are being reduced by greater use of inherently low-sulfur coals. Less than two decades after *The Clean Air Act* was passed, the sulfur content of coal purchased and burned by electric utilities had decreased 37 percent. Physically washing coal after it is mined and before it is burned is another way to reduce sulfur compound emissions. This process can remove sulfur-iron compounds (pyritic sulfur) from raw coal, but cannot remove organic sulfur, which is part of coal's molecular structure.

All these techniques represent a significant investment in maintaining clean air. A single scrubber, for example, can cost more than $100 million to construct, and many millions of dollars a year to operate. There are more than 140 scrubbers installed and operating at U.S. utilities, and about 50 more planned.

Electric utilities have already spent $60 billion to control sulfur emissions, and the investment, which continues every year, has paid off. According to the U.S. Environmental Protection Agency, sulfur dioxide emissions from electric utilities have gone down 18 percent from their 1973 peak. Taking into account that coal use by utilities has gone up dramatically during the period makes the accomplishment all the more impressive. As new, technologically advanced power plants replace older, less efficient ones, the trend toward lower sulfur dioxide emissions is expected to be enhanced.

As a fossil fuel, coal also contains carbon, which combines with oxygen to form carbon dioxide (CO_2) during the combustion process. CO_2 is one of five major so-called "greenhouse" gases, which help trap radiated heat back to the earth's surface.

This "greenhouse effect" is a natural process which maintains the earth's temperature at a level sufficient to support life. However, recent scientific and political debate has intensified over whether human activity — such as fossil fuel use and deforestation — has caused an acceleration of the natural greenhouse effect.

While most scientists agree that global atmospheric CO_2 and other greenhouse gases have risen in quantity over the past century, great disagreement remains over whether these increases have already — or will ever — affect the earth's climate.

Given coal's vital present and future role in meeting the world's energy needs, solutions to concerns over possible climate change will have to be global in nature and carefully balance environmental objectives with viable options for continuing to fully utilize fossil fuels.

New Technologies for Coal Combustion

The Clean Air Act, which has been in effect since 1970, and was last amended in 1990, is the most stringent air pollution control law in the world. Because of it, and through the combined efforts of business and industry, citizens and government, we enjoy some of the cleanest air to be found anywhere on the globe.

American industry has spent some $350 billion since 1970 to clean up the air, and each year the tab for pollution control totals another $33 billion.

Pollution control equipment accounts for up to 40 percent of the cost of a new power plant and 35 percent of operational costs, according to the Electric Power Research Institute. Those costs plus operating costs currently account for about $10 billion of the nation's electric bills each year, and will rise even higher under new Clean Air Act requirements.

Although environmental concerns about coal use center on the emission of sulfur and nitrogen compounds and carbon dioxide, coal is not the only for the leading source of them in our environment. However, coal combustion is a significant contributor, so as part of a national commitment to further reduce air pollution, more than a dozen advanced technologies for burning coal cleanly and more efficiently are being investigated. The development and demonstration of these technologies has required a substantial investment of more than $6 billion by government and private industry. Two of the leading technologies are:

Fluidized-bed Combustion (FBC)

Crushed coal mixed with limestone is supported on a strong rising current of air. The "fluidized" mixture acts as a boiling liquid, mixing turbulently, thus assuring efficient combustion. The limestone reacts with and removes over 90 percent of the sulfur. As a consequence, "scrubbers" (flue gas desulfurization) are not required for SO_2 control. Because the operating temperature is lower than in conventional boilers, the formation of nitrogen oxides is minimized.

The FBC technology lends itself to the design of smaller boilers that can be prefabricated as modular units, and at less capital cost than conventional boilers of the same generating capacity. Because of this savings, and the favorable economics of adding additional generating capacity to a power plant only as it is needed, electric utilities and consumers both are expected to find this technology attractive.

Coal Gasification

As an alternative to direct combustion of coal, in which the heat produced is used to develop steam to drive generator turbines, a great deal of progress has been achieved on technologies that depend on first gasifying the coal. The gas itself can be burned to power gas turbines, and then the remaining heat can be harnessed to produce steam to turn turbines.

This type of arrangement, called combined cycle gasification, is extremely efficient and clean. At one demonstration plant in California, emissions of combustion products were comparable to those from a natural gas-fired facility, allowing the plant to meet California's clean air requirements, the strictest in the U. S.. Another advantage of the coal gasification process is that it can be carried on in close proximity to the mine site. Rather than shipping the bulky coal long distances to a power plant, the power of coal can be shipped by wire from the gasification plant.

These and other advanced clean coal technologies variously hold the promise of generating more power with less fuel, and reduced operating and maintenance costs, greater pollution control; some produce marketable by-products; and many utilize smaller plants that can be modularly built. But they all speak to the same goal: making more effective and efficient use of an abundant energy resource.

Additional Words and Phrases

mine field：煤田
mudstone：泥岩
normal fault：正常断层
oblique fault：斜断层
outcrop：露头
possible ore：远景储量
sandstone：砂岩
coal mine resources development and utilization plan：煤炭资源开发利用方案
description：说明书
revised：修改地
location：（矿井）位置
mining method：采煤方法
coal seam：煤层
coal thickness：煤厚
strike length：走向长度
incline width：倾向宽度
mining height：开采高度（采高）
general situations：概况
overview：概述
traffic and location：交通位置
natural geography：自然地理

general situations of mineral resources：矿产资源概况
minefield structure and geological feature：井田构造及地质特征
mining conditions and hydrogeology conditions：开采技术条件及水文地质条件
mineral resources reserves：矿产资源储量
comments on geological survey report：对地质勘探报告的评述
construction plan：建设方案
mining plan：开采方案
water prevention and control program：水防治方案
prevention of the gas and coal dust explosion：防治瓦斯、煤尘爆炸
mine fire prevention：矿井防灭火
minefield development and mining：井田开拓与开采
mine characteristic and development principle：矿井特点及开发原则
minefield development：井田开拓
mining subsidence：开采沉陷
mining sequence：开采程序
mining-induced environmental damage：开采损害

Unit 2　A Kaleidoscope of Coal Mines

Foreword

In this unit, a kaleidoscope of coal mines will be introduced. As we all know, at present, all over the world, coal is regarded as one of the main energy sources. However, the coal is more than the resource of energy. Have you ever imagined that coal might have something to do with UFO? Have you ever known the ways to reduce SO_2 and the application of CCPs? You will enter a rather fantastic world where more facts about the coal will be explored.

Section One

Preview

Text A tells us that the things found in coal or mines may be the proof of alien life. The text mainly deals with the objects found in Russian in 2013 and 2015. In the text, the objects' shape, the objects' properties that puzzled the scientists, the scientists' reaction to the objects and so on are explored. Through the text, you will get to know another aspect of the coal.

Text A

Proof of Alien Life?

Nowadays, finding a strange artifact in coal is a relatively frequent occurrence. The first discovery of this sort was made in 1851 when the workers in one of the Massachusetts mines **extracted** a **zinc** silver-**incrusted** vase from a block of unmined coal which dated all the way back to the **Cambrian** era which was approximately 500 million years ago. Sixty one years later, American scientists from Oklahoma discovered an iron pot which was

Unit 2 A Kaleidoscope of Coal Mines

pressed into a piece of coal aged 312 million years old. Then, in 1974, an **aluminum** assembly part of unknown origin was found in a sandstone quarry in Romania. Reminiscent of a hammer or a support leg of a spacecraft "Apollo", the piece dated back to the **Jurassic** era and could not have been manufactured by a human. All of these discoveries not only puzzled the experts but also undermined the most fundamental **doctrines** of modern science.

The metal detail which was found by Vladivostok resident in 2013 is yet another discovery which perplexed the scientists. According to Komsomolskaya Pravda, a resident of Vladivostok — near the borders of China and North Korea — named Dmitry, noticed something odd about a hunk of coal he had obtained to heat his home during the winter. The coal in which the metal object was pressed was delivered to Primorye from Chernogorodskiy mines of Khakasia region. Knowing that the coal deposits of this region date 300 million years back, Russian experts inferred that the metal detail found in these deposits must be an **age-mate** of the coal.

When geologists broke the piece of coal in which the metal object was pressed into and spot-treated in with special chemical agents, it turned out that the metal detail was unusually light and soft. No more than seven centimeters long, the object was found to be composed of 98 percent aluminum and 2 percent **magnesium**. On the one hand, such an **alloy** stalled the scientists because nearly pure aluminum is very rarely found in nature. Thus, the detail was most definitely created artificially. On the other hand, however, when it became clear that the object was made from aluminum-magnesium alloy the experts quickly found an answer to the question of how a metal detail could withstand the ravages of time so well. The scientists explained that pure aluminum is increasingly prone to **oxidization** which contributes to the creation of a special layer protecting it from further **corrosion**. As a result, the metal detail made 98 percent from aluminum can endure not only high pressure but also heat and other severe natural conditions.

Another question that interests Russian scientists is whether the aluminum alloy is of Earthly origin. It is known from the study of **meteorites** that there exists **extra-terrestrial** aluminum-26 which subsequently breaks down to magnesium-26. The presence of 2 percent of magnesium in the alloy might well point to the alien origin of the aluminum detail. Nonetheless, further testing is needed to confirm this **hypothesis**.

The last property of the object that puzzled the scientists was its distinctive shape which was reminiscent of a modern **tooth-wheel**. It is hard to imagine that an object could take regular shape of a tooth-wheel with six identical "teeth" naturally. Moreover, the **intervals** between the "teeth" of the gear are curiously large in relation to the size of the "teeth" themselves which might mean that the detail was a part of a complicated mechanism. Nowadays, such "spare parts" are used in construction of microscopes and other mechanical appliances. This poses yet another unanswerable question to the modern scientists: how can the metal tooth-wheel be 300 million years old if the regular-shaped "wheel" itself was created by man millions of years later.

After the discovery came public, **conspirators** were quick to dub it "a UFO tooth-wheel". Russian scientists, however, do not jump to conclusions and will run further tests to learn more about the strange artifact.

In 2015, the bizarre flying saucer-shaped object was dug up during **excavation** work by a coal mining company in Siberia's Kuznetsk Basin, Russia, which aroused people's attention. The unusual object is almost perfectly circular, with a **diameter** of around 1.2 meters and weighing roughly 440 lbs — that's 31 stone.

Archaeologists were called in to examine the "craft", which the excavators believe is a man-made object, describing it as a "mystery". However, sky-watchers are convinced it fell from space.

Excavator Boris Glazkov, 40, who found the object, said: "I have to say it wasn't hard to see as it was really distinctive and large." "I've never seen anything like this object, which is obviously man-made out, here in the middle of nowhere before. It is a real mystery."

His colleague, Arthur Presnyakov, 38, said: "There were actually two similar objects, but the first one broke as it was being pulled out of the ground by the excavator bucket." "We thought we saw something sure, because it broke into pieces, but then when the second one appeared we stopped work and carefully removed it from the bucket."

Given that it was **embedded** so deep underground, it could be older than **mammoth** bones, which have been discovered in the area at a depth of 25 meters.

Unit 2　A Kaleidoscope of Coal Mines

But web-users have their own ideas. One of them commented: "It doesn't look man-made to me. Why on earth would a human create such a weird object and bury it so deeply? I think it must have come from outer space."

Scientists declined to comment further while the investigation is ongoing.

Vocabulary

extract / ɪkˈstrækt / *v.* to obtain a substance from something else using a particular process 提取；获得

zinc / zɪŋk / *n.* a bluish-white metal that is mixed with copper to produce brass and is often used to cover other metals to prevent them from rusting 锌

incrust / ɪnˈkrʌst / *v.* to cover or overlay with or as with a crust or hard coating 被……包住

aluminum / əˈluːmɪnəm / *n.* a silver-white metal that is light in weight, used, for example, for making cooking equipment and aircraft parts 铝

doctrine / ˈdɒktrɪn / *n.* a set of principles or beliefs, especially religious ones 教条；原则；教旨

age-mate / ˈeɪdʒmeɪt / *n.* person of the same or similar age 同龄人，年龄相仿的伙伴，年龄几乎相同的人（或动物）

magnesium / mæɡˈniːzɪəm / *n.* a light, silver-white metal that burns with a bright white flame 镁

alloy / ˈælɔɪ / *n.* a metal that is formed by mixing two types of metal together, or by mixing metal with another substance 合金

oxidization / ˌɒksədaɪˈzeɪʃən / *n.* the process of oxidizing or the addition of oxygen to a compound with a loss of electrons, which always occurs accompanied by reduction 氧化作用

corrosion / kəˈrəʊʒ(ə)n / *n.* a process by which something, especially a metal, is destroyed progressively by chemical action, as iron is when it rusts 腐蚀

meteorite / ˈmiːtɪəraɪt / *n.* stony or metallic object that is the remains of a meteoroid that has reached the earth's surface 陨石

extra-terrestrial / təˈrestrɪəl / *adj.* coming from outside the earth 地下的

hypothesis / haɪˈpɒθəsɪs / *n.* an idea or explanation of something that is based on a few known facts but that has not yet been proved to be true or correct 假说；前提；假定

tooth-wheel *n.* 齿轮

interval / ˈɪntəvəl / *n.* a period of time between two events 间隔；区间

conspirator / kənˈspaɪrətə / *n.* a person who is involved in a conspiracy 阴谋家

excavation / ˌekskəˈveɪʃ(ə)n / *n.* the activity of digging in the ground to look for old buildings or objects that have been buried for a long time 挖掘

diameter / daɪˈæmətər / *n.* a straight line going from one side of a circle or any other round object to the other side, passing through the center 直径

excavator / ˈekskəˌveɪtə(r) / *n.* someone who digs to find things that have been buried under the ground for a long time 发掘者，开凿者

embed / ɪmˈbed / *v.* to fix sth. firmly into a substance or solid object 嵌入式；内嵌；嵌入

mammoth / ˈmæməθ / *n.* an animal like an elephant larger than the mordern one and covered with hair, which lived on earth during the early stages of human development 猛犸

Phrases and Expressions

support leg：支撑腿；支柱
spare part：零件；配件；零配件
assembly part：装配部件；装配零件
sandstone quarry / ˈkwɒri / ：砂岩采石场
a hunk of：一块
coal deposits：煤藏
prone to：倾向于；易于……的
break down：分解

Proper Words and Terms

Cambrian / ˈkæmbrɪən /　寒武纪的；寒武纪
Oklahoma / ˌəʊkləˈhəʊmə /　俄克拉荷马州（美国）
Jurassic / dʒəˈræsɪk /　侏罗纪的；侏罗纪时代
Vladivostok / ˌvlædɪˈvɒstɒk /　海参崴（俄罗斯）

Unit 2 A Kaleidoscope of Coal Mines

Exercises

I. Comprehension of the Text

Directions: Please answer the following questions according to the text.
1. According to the text, in 1851, what was found by the workers?
2. What's the composition of the object discovered in the coal in 2013?
3. Why was the object found in 2013 believed to be made artificially?
4. According to the scientists, what's the reason that makes a metal detail withstand the ravages of time so well?
5. What fact made scientist doubt that the aluminum alloy is of Earth origin?
6. What's the shape of the object found in mine in 2015?
7. What made some people believe the object found in 2015 is from the outer space?
8. What was the Russian scientists' reaction to the objects found in the coal and the mine?

II. Group Discussion

Directions: Discuss the following questions with your partners, using as much text information as possible in your discussion.
1. Besides the things mentioned in the text, have you ever heard any other things found in the coal or some interesting stories related to coal?
2. Why do some people strongly believe the things found in the coal or mine are the proof of the alien life?

III. Word Bank

Directions: Fill the blanks with the words on the right side of the text. For each word, you can use only once.

The discoveries in coal or mines not only puzzled the experts but also undermined the most fundamental 1. _____ of modern science.

In 2013, a rail-shaped metal detail pressed in one of the pieces of coal was found in Russia, which perplexed the scientists. The object was found to be composed of 98 percent 2. _____, which is 3. _____ corrosion and 2 percent 4. _____. On the one hand, the

a) embedded
b) extracted
c) incrust
d) intervals
e) aluminum
f) excavator
g) tooth-wheel
h) prone to
i) comment

· 23 ·

Russian scientists believed it was the composition that prevented the metal detail from 5. _____. On the other hand, the presence of 2 percent of the metal in the alloy might well point to the alien origin of the aluminum detail. The metal detail's distinctive shape which was reminiscent of a modern 6. _____ also puzzled the scientists.

  In 2015, a flying saucer-shaped object, 7. _____ at a depth of around 40 meters was dug up during 8. _____ work by a coal mining company in Siberia's Kuznetsk Basin, Russia, which could be older than 9. _____ bones. Therefore, some people believed that it came from outer space.

  However, the Russian scientists treated the objects with caution. They refused to 10. _____ and did not jump to conclusions before the further studies.

j) mammoth
k) in relation to
l) magnesium
m) oxidization
n) doctrines
o) excavation

Ⅳ. Translation Practice

Directions: Translate the following sentences from Chinese to English.

1. 除了涉及温室效应以外,矿物燃料的燃烧也释放出氮氧化物、硫黄、烟尘及碳氢化合物等有害物质,造成了空气污染。(in relation to)
2. 众所周知,发电厂无论燃烧石油、煤或核燃料,都容易产生热量释放的问题。(be prone to)
3. 大部分地球上开采出的金属和燃料在被利用、废弃之后会被自然生态系统吸收。(extract)
4. 中国有文字记载的开采利用煤炭的历史可追溯到战国时代。(date back to)
5. 电磁发电机可将燃料燃烧产生的热能转化为电能。(convert ... to)
6. 对于煤矿中发现的飞碟形状物体是否证明了外星生命的存在,科学家们没有妄下结论。(jump to conclusion)

Ⅴ. Writing Practice

Directions: Write a 3-paragraph passage of about 120 words with the title "Coal and Its Application" based on the outline below.

1. Coal is the main energy source.
2. More than the energy source, coal is widely used in our daily life.
3. Coal and the development of the society.

Section Two

Preview

Text B deals with coal combustion byproducts. What are the components of CCPs? What are the applications of CCPs? How to reduce SO_2 emissions for a majority of electric power plants which generate electricity by burning coal? What is FGD? How about FGD technology and methods? Through the text, you will find the answers to all these questions.

Text B

Coal Combustion Byproducts

Electricity accounts for about 35% of the primary energy use in the United States and is produced by electric power **generators** designed to convert different fuel types into electricity. More than one-half the electricity in the United States is generated by burning coal, as a result, more than 100 million tons of solid **residues** known as coal combustion products (CCPs) are produced by electric utilities annually.

The majority of electric power utilities use high sulfur **bituminous** coal. Increased use of high-sulfur coal has contributed to an acid rain problem in North America. To effectively address this problem, the U. S. Congress passed the *Clean Air Act Amendments* of 1990 (CAAA'90-Public Law 101-549) with **stringent** restriction on sulfur oxides emissions.

CCPs are the resultant solid residues produced upon coal burning in the production of electricity. The coal is crushed, **pulverized**, and blown into a combustion chamber, where it immediately **ignites** to heat the boiler tubes. Upon burning of the coal, the inorganic **impurities**, known as coal ash, either remain in the combustion chamber or are carried away via the flue gas stream. Coarse ash particles (bottom ash and boiler slags) fall to the bottom of the combustion chamber, while the fine portion (fly ash) remains suspended in the flue gas. Before leaving the stack, these fly ash particles are removed from the flue gas by electrostatic precipitators or other available **scrubbing** systems, such as mechanical dust collectors, often called **cyclones**. Added to the ash is the FGD (flue gas desulfurization) material as the flue gas is "scrubbed" to remove the sulfur oxides.

The SO_2 reduction provision of CAAA'90, with its two phase implementation plan, has forced the electric utilities to find ways of reducing SO_2 emissions. Many utilities have switched to low-sulfur coal or fuel oil as partial and temporary solution to the problem. A significant number of powerplants still using medium- or high-sulfur coal installed flue gas desulfurization equipment from among a number of commercially available units. Wet lime FGD systems, later described, most commonly used in the United States, solve the SO_2 problem, but yield a large quantities of byproduct known as FGD material.

In addition to the FGD material, CCPs include fly ash, bottom ash, and boiler slag. Boiler slags represent a minor component of CCPs. One hundred percent of boiler slags is profitably utilized, mostly in the manufacture of **abrasives** and roofing **granules**. Among the major CCP components, bottom ash has over the years represented the highest use rate at approximately 35% of the amount produced.

FGD Technology

Numerous processes and equipment to reduce SO_2 and Nitrogen oxides (NOx) emissions have been developed, and many are commercially available. A significant number of electric power plants, which continue to use medium- and high-sulfur coal as fuel, have installed FGD equipment. The FGD systems are divided into two major types, wet systems (calcium, **sodium**, **ammonia**, magnesium, **potassium**, organic, and others) and dry systems (**reagent**, carbon **sorption**, combustion, metal oxide sorption, **catalytic** oxidation, SO_2 reduction, etc.).

FGD Methods

Passage of the CAAA'90 by the 101st Congress brought about FGD requirements for coal-fired power plants. This generated a **beehive** of activity in the research and development of FGD processes for the control of SO_2 emissions in the flue gas.

The major process division is between wet and dry systems. Wet systems completely **saturate** the flue gas with moisture while dry systems do not. With wet FGD systems, the fly ash is usually removed from the flue gas stream prior to the flue gas entering the FGD system. A large majority of the processes are systems that have been proposed but never carried beyond the laboratory due to poor performance or unfavorable economics.

Lime-Limestone-Based Systems

The great majority of FGD systems — about 90% — being installed in the United States utilize limestone or lime as a sorbent. Currently, more than 10,000 megawatts (MW) of electric power generating capacity use FGD systems. An additional 23,000 MW are either in the construction or the planning stages for **retrofit** with FGD units. Lime is more reactive than limestone; consequently, higher efficiencies can be obtained with lime as the sorbent, and thus lesser amounts are needed. This, in turn, results in reduced quantities of byproducts. In FGD systems using the quick lime (CaO) process, quick lime is slaked on site to form a calcium hydroxide **slurry**. This slurry is added to the reaction tank or scrubber basin, where it is mixed with recycled slurry from the scrubber. In the wet systems, the flue gas stream is sprayed with the lime slurry. The lime reacts with sulfur gases to form calcium **sulfite** and calcium sulfate. Sulfites formed need to be converted to sulfate. This is accomplished by increasing the oxygen content in the system, thus effecting the oxidation of sulfite to sulfate. The oxidation of sulfite to sulfate is dependent upon many process variables such as equipment design, PH, $O_2 : SO_2$ ratio, etc., and has caused serious operating problems due to scale formation in some systems.

Briefly, in a typical system, the flue gas comes in contact with a slurry of lime or limestone, ground to approximately 90% 200 mesh, and the reagent. The gas may or may not contain fly ash, depending on the absorber design. A larger amount of slurry (relative to the gas volume) is sprayed or dispersed in the contactor, saturating the flue gas and removing the SO_2. The scrubbed gas is then passed through mist eliminators and is often reheated to restore buoyancy prior to discharge.

The SO_2-rich liquor typically drains into large tanks where **neutralization** and precipitation reactions occur. Alkaline reagents may be added to the system to control the acidity.

Consumption

Components of CCPs have different uses as they exhibit distinct chemical and physical properties making each one suitable for a particular application. Current uses of CCPs include in cement, concrete, mine backfill, agriculture, blasting grit and roofing applications. Other current uses include, to a lesser extent, waste stabilization, road base-sub-base, and wallboard production.

Fly ash is used in the largest quantities accounting for 60% of the total CCPs used. Use in cement and concrete production tops the list of leading fly ash applications with more than 50%, followed by structural fills and waste stabilization. Approximately one-half of bottom ash was used in road base-sub-base, cement and concrete, structural fill, waste stabilization, and snow and ice control. Miscellaneous other applications, such as mineral fillers and extenders, flowable fill, etc., make up the other half of the use categories. Mining applications (about one-half of the total used), agriculture (about one-third), and blasting / roofing granules account for the bulk of FGD product uses, amounting to more than 90% of its total use. Virtually 100% of the boiler slags produced are utilized. Owing to its considerable **abrasivity**, boiler slag is used almost exclusively in the manufacture of blasting **grit** and roofing granules.

Vocabulary

generator / ˈdʒenəreɪtə(r) / *n.* a machine for producing electricity 发电机; a machine for producing a particular substance 发生器; a company that produces electricity to sell to the public 电力公司

residue / ˈrezɪdjuː / *n.* a small amount of sth that remains at the end of a process 残渣;残余物;残基

bituminous / bɪˈtjuːmɪnəs / *adj.* resembling or containing bitumen 沥青的,含沥青的

stringent / ˈstrɪndʒ(ə)nt / *adj.* very strict and that must be obeyed 严格的,严厉的; difficult and very strictly controlled because there is not much money 紧

缩的；短缺的

pulverize / ˈpʌlvəraɪz / *v.* to crush something into a fine powder 把……磨成粉状；to defeat or destroy somebody or something completely 粉碎（言论等）
ignite / ɪgˈnaɪt / *v.* to make something start to burn 点燃；着火
impurity / ɪmˈpjʊərəti / *n.* a substance that is present in small amounts in another substance, making it dirty or of poor quality 杂质
scrub / skrʌb/ *v.* clean with hard rubbling 擦洗；wash thoroughly 使净化
cyclone / ˈsaɪkləʊn / *n.* a violent tropical storm in which strong winds move in a circle 旋风；飓风
abrasive / əbˈreɪsɪv / *n.* a substance used for cleaning surfaces or for making them smooth（用来擦洗表面或使表面光滑的）磨料；研磨剂；研磨材料；磨具
granule / ˈgrænjuːl / *n.* a small, hard piece of something; a small grain 颗粒；颗粒剂
sodium / ˈsəʊdɪəm / *n.* a soft silver-white metal that is found naturally only in compounds, such as salt 钠
ammonia / əˈməʊnɪə / *n.* a gas with a strong smell 氨
potassium / pəˈtæsɪəm / *n.* a soft silver-white metal that exists mainly in compounds which are used in industry and farming 钾
reagent / rɪˈeɪdʒənt / *n.* a chemical agent for use in chemical reaction 试剂，反应物
sorption / ˈsɔːpʃ(ə)n / *n.* the taking in or holding of something, either by absorption or adsorption 吸附；吸着（作用）
catalytic / kætəˈlɪtɪk / *adj.* relating to or causing or involving catalysis 催化的
beehive / ˈbiːhaɪv / *n.* a structure housing a colony of bees 蜂窝；a hairstyle for women, with the hair piled high on top of the head 蜂窝状发型；any workplace where people are very busy 忙碌、拥挤的地方
saturate / ˈsætʃəreɪt / *v.* to make something completely wet 浸透，使湿透；to fill something or somebody completely with something so that it is impossible or useless to add any more 使充满；使饱和
retrofit / ˈretrəʊfɪt / *v.* to put a new piece of equipment into a machine that did not have it when it was built; to provide a machine with a new part, etc. 改进；翻新改造；*n.* the act of adding a component or accessory to sth. that did not have it while it was manufactured 更新，翻新，改造
slurry / ˈslʌrɪ / *n.* a thick liquid consisting of water mixed with animal waste,

clay, coal dust or cement 泥浆

sulfite / ˈsʌlfaɪt / **n.** a salt or ester of sulfurous acid 亚硫酸盐；亚硫酸盐或酯；硫化物

neutralization / ˌnjuːtrəlaɪˈzeɪʃn / **n.** the function or state of being neutralized 中和作用；中和反应

abrasivity / əbreɪˈsɪvɪtɪ / **n.** the quality of being abrasive 冲蚀度；磨蚀度

grit / ɡrɪt / **n.** very small pieces of stone or sand 砂砾，粗砂

Phrases and Expressions

convert into：转换成；转化成；转换为
bituminous coal：烟煤
combustion chamber：（发动机等的）燃烧室
boiler tube：锅炉水管墙
bottom ash：炉底灰；残渣；底渣；炉渣
boiler slag：锅炉渣
electrostatic precipitator：静电除尘器；静电沉淀器
flue gas desulfurization：烟气脱硫
prior to：在……之前；先于
quick lime：生石灰
calcium hydroxide：氢氧化钙；熟石灰
reaction tank：反应槽；反应池；反应罐
calcium sulfate：硫酸钙
scale formation：结垢生成；无机盐结垢
precipitation reaction：沉淀滴定反应；析出；沉淀反应
alkaline reagent：碱性试剂
mine backfill：矿山填充，回填
blasting grit：喷砂
roofing application：屋面铺设
mineral fillers：矿物填料
flowable fill：可流填充

Unit 2 A Kaleidoscope of Coal Mines

Exercises

Comprehension of the Text

Directions: Choose the right answer to each question according to the text.

1. According to the text, what's the aim of *Clean Air Act Amendments* of 1990?
   A. To ban the use of high-sulfur coal.
   B. To improve the ecological environment.
   C. To decrease the quantity of electric power plants.
   D. To punish the electric power plants.
2. Among the CCP components, _____ represents the highest use rate.
   A. fly ash
   B. bottom ash
   C. FGD
   D. boiler slag
3. According to the text, which of the following statements is true?
   A. In the United States, limestone or lime is utilized as a sorbent in the great majority of FGD systems.
   B. All the electric power plants have installed FGD equipment.
   C. It is universally recognized that FGD units are perfect, for they solve the SO_2 problem.
   D. The FGD systems are divided into a variety of types.
4. Whether the gas contains fly ash depends on _____.
   A. the characteristic of the fly ash
   B. the usage of limestone
   C. the usage of lime
   D. the absorber design
5. According to the text, components of CCPs could be used in various fields because of _____.
   A. their quantity
   B. FGD system
   C. their chemical and physical properties
   D. advanced technology

Section Three

Extended Reading

Interesting Facts about Coal

Coal was formed about 300 million years ago.

Coal is a combustible mostly black sedimentary rock composed mostly of carbon and hydrocarbons.

Coal takes a million years to create and therefore it belongs to non-renewable energy sources.

Coal mining uses two methods: surface or underground mining where surface mining is more dominant method because it is less expensive than the underground mining.

Coal is mostly transported by train.

Coal as well as the other fossil fuels isn't ecologically acceptable because of CO_2 and global warming.

Coal is classified into four main types: lignite, subbituminous, bituminous and anthracite, and the coal value is determined by the amount of the carbon it contains.

Coal is mined in 27 US states.

Coal is mainly used for generating electricity (more than 90% of US coal).

Coal usually has a negative impact on environment. Mining can damage ground and surface waters and when coal burns as the fuel it releases CO_2, which is the main greenhouse gas that causes global warming.

Coal is so called "dirty" energy source because of its negative effect on environment.

Coal could be the world's most attractive fuel in years to come thanks to the methods of coal purification which are resulting in cleaner coal, removing sulfur and other dangerous elements.

Coal is used on the large scale in China and USA.

Coal can be answer for future only if technology will enable "ultra-clean"

Unit 2 A Kaleidoscope of Coal Mines

coal.

Coal must be relatively dry before it can be burned successfully.

World coal consumption is more than 5.3 billion tons annually, of which three quarters are used for generating electricity.

Coal was already used in the Bronze age (Britain).

Coal's share in the total world electricity production is about 40%.

Coal deposits could be enough to satisfy current world energy needs for the next 300 years.

Coal is getting more attention because of the increased prices of oil and natural gas.

Coal can be converted into gasoline or diesel by a couple of different processes, for instance, the Fischer-Tropsch process, Bergius process and Karrick process.

Coal is the official state mineral of the Kentucky and the official state rock of Utah.

Coal total reserves are approximately about 998 billion tons.

Coal is mined in more than 100 countries.

Coal is the main reason for China's economic boost and for China's environmental problems as well.

Coal is a combustible mineral formed from the remains of trees, ferns and other decayed plants that existed and died up to 400 million years ago in some cases.

Coal has been used as an energy source for hundreds of years and was part of international trade in as long ago as the Roman Empire.

Coal provided the energy which fueled the Industrial Revolution of the 19th Century and also launched the electric era in the 20th Century.

37% of the electricity generated worldwide is produced from coal.

Coal is by far the cheapest source of power fuel per million BTU, averaging less than half the price of petroleum and natural gas.

The world's iron and steel industry depends on the use of coal.

The value of coal produced in the United States each year is nearly $20 billion.

Coal is directly responsible for the existence of more than 90,000 U.S. jobs and nearly one million jobs directly and indirectly.

Coal mining has a combined direct and indirect impact of $161 billion annually on the U.S. economy. This is $596 for every U.S. citizen.

The federal government of U.S. receives more than $11 billion annually in taxes and fees from the coal industry.

State and local governments receive nearly $9 billion each year in revenues.

Developing countries' demand for coal will double through 2020, according to the Energy Information Administration (EIA).

Coal reserves are spread over almost 100 countries. Proven coal reserves are estimated to last over 200 years with current production rates. In contrast, proven oil and gas reserves are equivalent to around 40 and 60 years.

America's coal is used primarily for the production of electricity.

There were 315,000 Megawatts (net) of coal-based electrical generating capacity in the United States.

Some 965 million tons of coal were consumed for the generation of electricity. This amounted to 86% of total U.S. coal production.

Many large countries contain significant proven reserves of coal. While data quality varies widely, the countries with the greatest estimated recoverable reserves of coal are:

United States: 273 billion tons.

Russia: 173 billion tons.

China: 126 billion tons.

India: 93 billion tons.

Australia: 90 billion tons.

In U.S. coal was first discovered in what is now West Virginia in 1742 by John Peter Salley in what is now Boone County.

McDowell County has produced more coal than any other county in West Virginia.

The coal industry pays approximately $70 million in property taxes annually in U.S..

The Coal Severance Tax adds approximately $214 million into West Virginia's economy.

Twenty-four million dollars of coal severance taxes collected each year goes directly into the Infrastructure Bond Fund.

All 55 counties, even the non-coal producing counties receive Coal Severance Tax funds.

The coal industry payroll is nearly $2 billion per year.

Coal is responsible for more than $3.5 billion annually in the gross state product.

The United States produces about 20%, or 1.1 billion tons, of the world's coal supply — second only to China.

The United States has about a 245-year supply of coal, if it continues using coal at the same rate at which it uses coal today.

Montana is the state with the most coal reserves (119 billion tons). But Wyoming is the top coal-producing state — it produced about 400 million tons in 2004.

Texas is the top coal-consuming state. It uses about 100 million tons each year.

Coal ash, a byproduct of coal combustion, is used as filler for tennis rackets, golf balls, and linoleum.

U.S. coal deposits contain more energy than that of all the world's oil reserves.

Each person in the United States uses 3.8 tons of coal each year.

Coal is the world's most abundant fossil fuel — more plentiful than oil and natural gas. It is second only to oil as a source of energy that we use. Coal is widely used because it's easily obtained. There's lots of it. It's well distributed throughout the world, and it has a high energy content.

Coal has many important uses, but most significantly in electricity generation, steel and cement manufacture, and industrial process heating.

Coal provides over 23% of global primary energy needs. It generates around 39% of the world's electricity. Almost 70% of total global steel production is dependent on coal.

There is more stored energy in Canadian coal than all the country's oil, natural gas, and oil sands combined.

Canada exports about 28 million tons of coal annually to more than 20 countries.

Canada ranks tenth in the world in total coal reserves with 4 billion tons of bituminous coal. That's coal covering a football field to a height of about

4,500 kilometers!

Additional Words and Phrases

after treatment：后处理
aromatic：芳族化合物
as-received：收到基（煤）
atmospheric pollution：大气污染
auto-ignition temperature：自燃温度，自燃点
biodiesel：生物柴油
biomass：生物质
carbon balance：碳平衡
carbon(CO_2) capture and storage：回收并储存碳（二氧化碳）
catalyst：催化剂
coal slurry：水煤浆
coal steam-electric plant：火电厂
coal steam plant with FGD：火力发电厂烟气脱硫
cold start：冷态启动
cool water demonstration：冷水示范电厂
cooling water：冷却水
DCL (direct coal liquefaction)：煤直接液化
direct liquefaction technology：直接液化技术
electricity or power generation：发电
emission rate：排放率
fuel cycle：燃料循环
gasification：气化
GHG emissions mitigation：减排温室气体
ICL (indirect coal liquefaction)：煤间接液化
IGCC plant：整体煤气化联合循环电厂
mitigation：减少，减排
particulate matter：颗粒物
gas-phase reactor：气相反应器
global warming：全球变暖
GTL (gas to liquids)：气变油

HC fuel：烃类燃料
health cost：健康损害
liquid-phase：液相
liquid-phase reactor 液相反应器
location factor：区域因子
noxious material：有害物质
oxygenated fuel：氧化燃料
ozone：臭氧
power sector：电力行业
purge gas：净化气体
recycle design 循环方式
renewable energy：可再生能源
soot：烟灰
synthesis：合成
synthetic fuel：合成燃料
toxic metal：有毒金属物
vapor pressure：蒸汽压

Unit 3 Coal Industry and Chinese Social Development

Foreword

Unit 3 introduces the role of coal industry in Chinese social development. Text A tells about China's heavy dependence on coal in terms of economy and social life. Text B depicts a picture of China's energy reserves and supplies, in which coal plays and will continue to play a major role into the future. Text C probes into the future of coal industry. Given its unshakable lead over other resources, the importance of coal industry should never be underestimated so as to ensure the healthy and rapid development of Chinese society.

Section One

Preview

Text A tells about the importance of coal in China. Being the world's leading user and producer of coal, China depends heavily on coal both economically and socially. Even with other energy resources being exploited, coal provides 65% of China's energy needs for the foreseeable future. It's impossible to replace coal use for quite some time. Besides, Chinese family life quality goes hand in hand with coal industry, which produces low cost energy resources to heat homes and cook meals, and even offers job opportunities for coal briquette sellers and scavengers.

Text A

China's Dependence on Coal

China is the world's leading user and producer of coal. It burns 24 percent of the world's coal, compared to 25.5 percent in the United States and 7 percent in India. In 2006, over 2.2 billion metric tons of coal was taken from Chinese coal mines, "more than the United States, India and Russia combined". This was up from 2.1 billion in 2005, 1.7 billion in 2003 and 1.4 billion in 2002 and 70 percent more than 2000. The figure is expected to rise to 2.6 billion in 2010 and 3.1 billion in 2020, with coal production increasing at a rate of 10 percent to 15 percent a year.

Coal and Economy

Coal provides China with 69 percent of its primary energy and 80 percent of its electricity. It is relatively plentiful and cheap but very dirty. The government wants to replace coal with oil, natural gas and **hydroelectric** power primarily to clean up its air. Even so demand for coal keeps increasing as the economy grows and 60 new coal-fired power stations go on line every year. China could become a net coal importer between 2010 and 2015. China relies on coal to achieve its goal of energy independence. Thanks to coal China is 90 percent **self-sufficient** in energy.

Even if **drastic** measures are taken, China will still have to rely on coal for 65 percent of its energy needs for the **foreseeable** future. A Chinese coal engineer told the *New York Times*, "We are a developing country and we started without a very good foundation. We have so few choices and no funding, so our industries are going to rely heavily on coal for a long time to come." By some estimates China will have to **triple** its use of coal if it wants to achieve a standard of living near that of the United States. Some predict that by 2030 China's demand for coal will exceed that of the rest of the world combined. Even though China far and away produces more coal than any other country, demand is so strong that it has limited exports.

It is estimated that even with new land gas pipelines opening and **alterative** energies being exploited, China will get at least 50 percent of its energy from coal in 2020 and beyond. According to the World Resources Institute, "China has a lot of coal, has very limited supply of other fossil fuels and even with rapid growth rates in **renewables**, it will be difficult to actually replace the coal in use for quite some time." David Fridley, a China energy expert at the University of California, Berkeley, told the *New York Times*, "Only coal can provide new capacity in the time and scale needed."

Coal production has been slowed by **crackdowns** on illegal mines and

transportation bottlenecks that keep coal from reaching all the places it is needed. This has led to a rise of coal prices. The increase of coal prices in turn has **devastated** the electricity-generating industry because companies that produce electricity are tightly regulated by the government and are not allowed to raise prices without government approval. The government, fearing a **backlash** from consumers, doesn't want to raise prices. After coal prices **spiked** in 2003 and 2004, 85 percent of coal power plants lost money.

Coal and People's Life

Coal is used at a rate of about 1.5 tons per person a year. It is used to generate power for factories and electric plants as well as cook meals and heat homes. China burns the stuff so fast that railroads cannot deliver it fast enough and Chinese ships sit outside ports in Australia for weeks waiting for shipments of coal.

Jerry Goodell wrote in *Natural History* magazine: "In China coal is everywhere. It's piled up on sidewalks, pressed into bricks, and **stacked** near the back doors of homes. It's **stockpiled** into small mountains in open fields, and **carted** around behinds bicycles and wheezing **locomotives. Plumes** of coal smoke rise from **rusty** stacks on every urban horizon. **Soot** covers every **windowsill** and ruins the collar of every white shirt. The Chinese burn less coal per capita than America does, but in sheer tonnage, they burn twice as much."

Water and pulverized coal are formed into cakes and bricks at farms throughout China. Peasants and city dwellers use these cakes for cooking and heating. In many places coal bricks are still delivered to people's homes by tricycles or bicycles. One skinny 38-year-old coalman said, "I just ride around with no fixed destination, just wherever people need coal to heat their homes." Delivering coal is his winter job. He also works on a farm in the spring and summer.

Coal has enriched a few provided-hard low-paying jobs for many but brought misery to many more in the form of dirty air and water and **scarred landscapes**. Poor **scavengers rummage** through hill-size **slag** heaps for usable chunks of coal to heat their own homes or sell. Describing one, a reporter with National Geographic wrote, "High on one steep incline, nearly 500 feet up, a scavenger **trawls** for usable fragments of coal, **dodging** fresh loads of rock **careening** down the **embankment**, **sidestepping** the coal **embers smoldering** beneath the surface."

Vocabulary

hydroelectric / ˌhaɪdrəʊˈlektrɪk / *adj.* relating to or involving electricity made from the energy of running water 水力发电的

self-sufficient / ˈselfsəˈfɪʃənt / *adj.* able to do or produce everything that you need without the help of other people 自给自足的；自立的

drastic / ˈdræstɪk; ˈdrɑː- / *adj.* extreme in a way that has a sudden, serious or violent effect on something 极端的；急剧的；严厉的；猛烈的

foreseeable / fərˈsɪəb(ə)l / *adj.* that you can predict will happen; that can be foreseen 可预料的；可预见的；可预知的

triple / ˈtrɪp(ə)l / *adj.* having three parts or involving three people or groups 三部分的，三人的，三组的；three times as much or as many as something 三倍的，三重的；*v.* to become, or to make sth. three times as much or as many 使成为三倍，使增至三倍

alterative / ˈɔːltərətɪv / *adj.* likely or able to produce alteration （趋于）改变的；a drug that restores normal health 使体质回归正常的药品

renewable / rɪˈnuəb(ə)l / *n.* natural resources such as wind, water and sunlight which are always available 可再生能源；*adj.* that is replaced naturally or controlled carefully and can therefore be used without the risk of finishing it all 可更新的，可再生的，可恢复的；that can be made valid for a further period of time after it has finished 可延长有效期的，可展期的，可续订的

crackdown / ˈkrækdaʊn / *n.* severe action taken to restrict the activities of criminals or of people opposed to the government or somebody in authority 严厉的打击，镇压

bottleneck / ˈbɒtlnek / *n.* a narrow or busy section of road where the traffic often gets slower and stops 隘路；狭道（常引起交通阻塞）；anything that delays development or progress, particularly in business or industry（尤指工商业发展的）瓶颈，进展之阻碍，障碍

devastate / ˈdevəsteɪt / *v.* to completely destroy a place or an area 使荒废，摧毁，毁灭；to make somebody feel very shocked and sad（常用被动语态）使震惊，使极为忧伤，使极为悲痛

backlash / ˈbæklæʃ / **n.** a strong negative reaction by a large number of people, for example to something that has recently changed in society（对社会变动等的）强烈抵制，集体反对

spike / spaɪk / **v.** a sudden large increase in something 猛增；急升；**n.** a thin object with a sharp point, especially a pointed piece of metal, wood, etc. 尖状物，尖头，尖刺；a metal point attached to the sole of a sports shoe to prevent athletes from slipping while running（防滑）鞋钉

stack / stæk / **v.** arrange something in neat piles. （使）放成整齐的一叠（或一摞、一堆）

stockpile / ˈstakˌpaɪl; ˈstakpaɪ / **n.** a large supply of something that is kept to be used in the future if necessary 囤聚的物资，大量储备；**v.** to collect and keep a large supply of sth. 大量储备

cart / kɑː(r)t / **v.** to carry something in a cart or other vehicle 用马车运送；用车装运

locomotive / ˌləʊkəˈməʊtɪv / **n.** a railway engine that pulls a train 机车；火车头

plume / pluːm / **n.** a cloud of something that rises and curves upwards in the air 飘升之物

rusty / ˈrʌsti / **adj.** covered with rust 生锈的

soot / sʊt / **n.** black powder that is produced when wood, coal, etc. is burnt 煤烟；油烟

windowsill / ˈwɪndəʊsɪl / **n.** a narrow shelf below a window, either inside or outside 窗沿；窗台

scar / skɑː / **v.** to form or become marked by a scar 结疤，使留伤痕，（比喻）使留痕迹，弄丑；**n.** a mark left (usually on the skin) by the healing of injured tissue 疤

landscape / ˈlændˌskeɪp / **n.** everything you can see when you look across a large area of land, especially in the country（陆上，尤指乡村的）风景，景色

scavenger / ˈskævɪndʒə(r) / **n.** an animal, a bird or a person that scavenges 食腐肉的兽（或鸟）；捡破烂的人；拾荒者

rummage / ˈrʌmɪdʒ / **v.** to move things around carelessly while searching for something 翻寻；乱翻；搜寻

slag / slæg / **n.** the waste material that remains after metal has been removed from rock 矿渣；熔渣；炉渣

trawl / trɔːl / **v.** to search through a large amount of information or a large number of people, places, etc. looking for a particular thing or person 查阅（资料）；搜集，搜罗网罗（人或物）

dodge / dɒdʒ / *v.* to move quickly and suddenly to one side in order to avoid somebody or something 闪开，躲开，避开；to avoid doing something, especially in a dishonest way（尤指不诚实地）逃避

careen / kəˈriːn / *v.* to move forward very quickly especially in a way that is dangerous or uncontrolled（尤指危险或失控地）猛冲，疾驶

embankment / ɪmˈbæŋkmənt / *n.* a wall of stone or earth made to keep water back or to carry a road or railway over low ground 堤，堤岸，堤围；a slope made of earth or stone that rises up from either side of a road or railway（公路和铁路）路堤（公路或铁路两侧的）护坡

sidestep / ˈsaɪdstep / *v.* to avoid something, for example being hit, by stepping to one side 横跨一步躲过，侧移一步闪过；to avoid answering a question or dealing with a problem 回避，规避（问题等）

ember / ˈembər / *n.* a piece of wood or coal that is not burning but is still red and hot after a fire has died 余火未尽的木块（或煤块）

smolder / ˈsməʊldə(r) / *v.* to burn slowly without a flame（无明火地）阴燃，闷烧

Phrases and Expressions

go on line：（项目）持续增加
by estimates：据估计
far and away：……得多；远远
per capita：每人；人均；按人口计

Proper Words and Terms

The World Resource Institute: a non-governmental global research organization which seeks to create equity and prosperity through sustainable natural resource management. 世界资源研究所

University of California, Berkeley: a public research university located in Berkeley, California. It is the flagship campus of the University of California system, one of three parts in the state's public higher education plan, which also includes the California State University system and the California

Community Colleges system. 加州大学伯克利分校

New York Times: an American daily newspaper, founded and continuously published in New York City since September 18, 1851, by The New York Times Company. It has won 114 Pulitzer Prizes, more than any other news organization《纽约时报》

Natural History: a magazine published in the United States. The stated mission of the magazine is to promote public understanding and appreciation of nature and science.《博物学》杂志

National Geographic: formerly *The National Geographic Magazine*, is the official magazine of the National Geographic Society. It has been published continuously since its first issue in 1888, nine months after the Society itself was founded. It primarily contains articles about geography, history, and world culture. The magazine is known for its thick square-bound glossy format with a yellow rectangular border and its extensive use of dramatic photographs.《国家地理》杂志

Exercises

I. Comprehension of the Text

Directions: Please answer the following questions according to the text.

1. Among coal, oil, natural gas and hydroelectric power, which are clean energy resources?
2. Can gas and other alterative energy fully replace coal in China in the near future?
3. What factors slowed China's coal production?
4. What is the result of slowing coal production?
5. Can the present coal supply in China fully satisfy China's need?
6. What is coal used for in Chinese cities?
7. What are the negative effects brought about by coal?
8. Besides providing energy, what other benefits does coal bring?

II. Group Discussion

Directions: Discuss the following questions with your partners, using as much text information as possible in your discussion.

1. Can you use your own experiences to illustrate the social, economic, and

Unit 3 Coal Industry and Chinese Social Development

academic significance of coal in your life?
2. Given China's great demand for coal and the environmental disaster brought about by coal, do you have any idea as to how to enhance people's life standards and boost economic development without causing too much pollution?
3. Coal is in short supply in present China. So is it still worthwhile to crack down illegal mines? As far as you are concerned, what harm is done by illegal mines?

III. Word Bank

Directions: Fill the blanks with the words on the right side of the text. For each word, you can use only once.

China, with its 1. _____ economy and constantly enhanced living standards, 2. _____ heavily on coal. On the one hand, with its robust economic development, energy 3. _____ in China is huge. While 4. _____ energy resources are still under developing and 5. _____, at least 50% of its energy comes from coal in 2020 and beyond. Only coal can provide new capacity in the time and scale needed. Besides, coal production must be ensured so as not to bring money loss for 6. _____ companies because their costs are raised while electricity price is 7. _____ and suppressed by the government. On the other hand, coal is closely linked to Chinese life quality. City people use coal for cooking and 8. _____. Coal can also provide job opportunities for people like coal 9. _____ sellers and 10. _____. All in all, given the indispensable role played by coal in economic development and people's life, the consolidation of coal industry can never be slowed down for

a) exploitation
b) alterative
c) briquette
d) relies
e) electricity-generating
f) heating
g) stockpiled
h) scavengers
i) regulated
j) pulverize
k) boosting
l) self-sufficient
m) consumption
n) foreseeable
o) drastic

now and in the near future.

IV. Translation Practice

Directions: Translate the following sentences from Chinese to English.

1. 随着专门学校出身的新队员加入国家队,2018年金牌数有望突破20。(is expected to, with, academy)
2. 据估计,该项目的投入将超过2000亿元。(by some estimates, exceed)
3. 教师的武断很容易招致学生的叛逆,而学生的叛逆又导致教师的教学热情消退,因为他们没有得到学生的肯定反应。(assertiveness, lead to, in turn, hamper, because)
4. 每一个留守儿童的眼里都写满了失落;每一个留守妻子的脸上都蔓延着思念。(stay-at-home children, every … every … , loss)
5. 农民工的确比待在家里的农民挣得多多了,但是这些都是付出巨大代价换来的,他们感觉孤独,并且因为不能陪伴孩子和父母感到很内疚。(migrating farmers, but, at great costs, guilt, in the form of / because)
6. 城里的人看起来总是行色匆匆,商店里,他们焦急地寻求店员的服务,或者推搡他人,以便尽快结束购物。(appear to be hurrying, restlessly, seeking attention, elbowing)

V. Writing Practice

Directions: Write a 3-paragraph passage of about 120 words with the title "Coal Mining: Encourage or Curb" based on the outline below.

1. Some people take a hard line that more coal should be mined since coal serves as the motor for China's economy.
2. But others hold the view that coal mining must be limited given its negative impacts.
3. Confronted with the polarized attitudes, my stance is …

Section Two

Preview

Text B introduces China's energy reserves and supplies, in which coal plays and will continue to play a major role in the future. In energy reserves, coal accounts for 73.4% of the proven reserves of the conventional energy, far more important than other energy resources like oil and gas. The distribution of China's energy resources varies from region to region. In energy supplies, coal still has a commanding lead over other resources. And China's develoment and exploitation strategy is based on the different geographic locations of the resources. For example, "coal by wire" plan will be implemented to ensure electricity supplies along the coast as coal is rich in the north-east regions.

Text B

Coal in the Energy Reserves and Supplies of China

Energy Reserves

With coal accounting for 73.4% of the proven reserves of conventional energy in China and 94.3% of fossil energy, it is inevitable that this fuel will continue to play the major role in supplying China's energy needs in the future. Coal reserves at the end of 1997 were estimated at 114.5 billion tons or some 57.2 billion tons which, at current rates of production, would last for nearly a century. In comparison, oil reserves were estimated at 3.3 billion tons, with a production / reserves ration of 20.5 years, and gas reserves were estimated at 1.16 **trillion** cubic meters, or some 1.1 billion tons, with a production / reserves ration of 52 years.

The distribution of energy resources varies widely from region to region, with nearly 67% of coal reserves situated in north and north-western China, while oil and gas reserves are concentrated principally in the north-east, east and far west of China, and off-shore. Northern China is the most energy rich region, followed by the south-west, due to **hydro**, and the north-west.

Energy Supplies

Compared to 17.6% for oil, 2.1% for gas, 1.8% for hydro and 0.4% nuclear, coal is the **principal** fuel source, accounting for over 78% of **indigenous** energy production in 1996, with the principal deposits located in Shanxi province. Chinese coal production has more than doubled from 620 million tons in 1980 to nearly 1.4 billion tons in 1996, helping to fuel the country's spectacular economic growth.

China is the world's sixth largest oil producer and, in 1997, produced 3.2 million barrels of oil per day, primarily for internal consumption. Nearly 90%

is produced onshore and nearly one-third of the total production comes from the Daqing field in north-eastern China. Since 1993, China has become a net importer of crude oil and, in search of new reserves, China has been targeting the remote Tarim Basin in the north-west corner of the country and the **offshore** areas in the South China Sea.

While natural gas production in China has increased from 3.4 billion cubic meters in 1971 to 22.7 billion cubic meters in 1997, it still remains a **marginal** fuel within the Chinese energy balance and under-utilized relative to the total resource base. Most gas is produced onshore in Sichuan province in western central China, but further developments are being targeted both on and off shore. The largest offshore gas field, Yacheng, began production in 1996, supplying gas to Hainan Island and to a power plant in Hong Kong. The current five-year plan **foresees** an annual production of 25 billion cubic meters by the year 2000 and close to 30 billion cubic meters by the year 2005.

Electricity production more than doubled between 1987 and 1997 to 1,134 TWh, with coal-fired generation the overwhelming source of electricity in China. **Thermal** generation accounted for 81.3%, hydro 17.4% and nuclear 1.3% in 1996. Coal-fired generation alone accounted for 75%. A "Business-as-Usual" scenario estimates electricity generation growing by an average 5.4% per **annum** to reach 2,497 TWh by 2010 and 3,857 TWh by 2020.

China's strategy for energy development is based on three main geographical areas:

1) coal will be the main source of energy in the north-east regions of China,

2) nuclear power will be the main focus of energy development in the western regions,

3) hydro-power will provide most supplies of energy in the southern regions.

Installed electricity generating capacity in 1995 was 227 GW, of which 70% was coal-fired, 23% hydro, 6% oil-fired and the remainder split between gas-fired and nuclear. Annual capacity additions over the past six years have been some 16 GW. Installed capacity could reach 757 GW by the year 2020, with 62% coal-fired, 26% hydro, 6% oil and 3% nuclear. At least a fifth of the US $ 100 billion of investment required in the 1996 to 2000 period is projected to come from overseas.

China's current Ninth Five Year Plan (1996-2000) targets the development of "coal by wire", the construction of thermal power plants near mine-mouths with the electricity distributed to the main consumer demand centers on the coast by high voltage, long distance transmission lines. The 2,100 MW Yangcheng power station in Shanxi province is the first of such projects, with the power generated by this plant **destined** for Shanghai and the lower Yangtze industrial region. The first of the six 350 MW units is expected to enter service in mid-2000.

China's development of nuclear power is in its early stages, with the 300 MW **pressurized** water Qinshan plant, south of Shanghai, coming into commercial operation in 1993 and the 2 × 900 MW Daya Bay **complex** near Hong Kong in 1993. There are a further four plants under construction, and official plans call for 20 GW of nuclear capacity by the year 2010 and between 40 to 50 GW by 2020.

China has the largest hydroelectric potential in the world estimated at 675 GW, of which 290 GW is economically **exploitable** and 56 GW actually exploited (in 1996). By 2020, hydro-power capacity may reach almost 200 GW, including the 18.2 GW **Three Gorges** project on the Yangtze River.

Vocabulary

trillion / ˈtrɪljən / *n.* one million million（美）万亿,兆；one million million million（英）百万兆

hydro / ˈhaɪdrəʊ / *n.* short for hydroelectricity 水电

principal / ˈprɪnsəp(ə)l / *adj.* first in order of importance 最重要的,主要的

indigenous / ɪnˈdɪdʒɪnəs / *adj.* belonging to a particular place rather than coming to it from somewhere else 本土的；土著的；国产的；固有的

offshore / ˌɒfˈʃɔr / *adj.* happening or existing in the sea, not far from the land 海上的；近海的

marginal / ˈmɑːdʒɪn(ə)l / *adj.* small and not important 小的,微不足道的,不重要的；not part of a main or important group or situation 非主体的,边缘的

foresee / fɔːˈsiː / *v.* to think something is going to happen in the future; to know about something before it happens 预料；预见；预知

thermal / ˈθɜːm(ə)l / ***adj.*** connected with heat 热的；热量的

annum / ˈænəm / ***n.*** year (拉丁语)年，岁

destine / ˈdestɪn / ***v.*** (usu. pass.) set apart, decide or ordain in advance (通常用于被动语态)区分；指定；意欲；筹划

pressurize / ˈpreʃəraɪz / ***v.*** to increase the pressure of a gas or liquid in a container beyond normal levels 对……加压力

complex / kəmˈpleks / ***n.*** a group of buildings of a similar type together in one place (类型相似的)建筑群；a conceptual whole made up of complicated and related parts 相关联的一组事物

exploitable / ɪksˈplɔɪtəbl / ***adj.*** able to be used or developed 可开发的；可利用的

per annum：每年

Proper Words and Terms

Three Gorges：one of the biggest hydropower-complex project in the world, ranking as the key project for improvement and development of Yangtze River. The dam is located in the areas of Xilingxia gorge, one of the three gorges of the river, which will control a drainage area of 1 million km^2, with an average annual runoff of 451 billion m^3. The open valley at the dam site, with hard and complete granite as the bedrock, has provided the favorable topographical and geological conditions for dam construction.

Exercises

Comprehension of the Text

Directions: Choose the right answer to each question according to the text.

1. How long will coal production last with the current rate of production?
   A. nearly a century
   B. nearly half a century
   C. about 20 years
   D. 67 years

2. Where are most coal reserves situated in China?
   A. north east
   B. north and north west

C. far west

D. south west

3. In the sentence "it still remains a marginal fuel within the Chinese energy balance and under-utilized relative to the total resource base." in paragraph 5, what does "marginal" mean?

    A. principal

    B. important

    C. empty

    D. unimportant

4. Which is true among the following statements according to the text?

    A. Coal will be the main source of energy development in the western region.

    B. Solar power will be the main source of energy in the western region.

    C. Hydro-power will provide most energy supplies in the eastern region.

    D. Coal will be the main source of energy development in the north-east region.

5. Which statement is true about the development of "coal by wire" in the Ninth Five Year Plan?

    A. Thermal power plants will be constructed far from the mine-mouths.

    B. The electricity produced will be supplied to the west.

    C. Thermal power plants will be constructed near the mine-mouths.

    D. The electricity produced will be supplied to only Shanghai.

Section Three

Extended Reading

Transforming China's Grid — Will Coal Remain King in China's Energy Mix?

Coal has been the primary fuel behind China's economic growth over the last decade, growing 10 percent per year and providing three quarters of the nation's primary energy supply. Like China's economy, coal's use, sale and broader impacts are also dynamic, changing with technology and spurring policy interventions. Currently, China's coal sector from mine to boiler is undergoing a massive consolidation designed to increase efficiency. Coal's supreme position in the energy mix appears to be unassailable.

However, scratch deeper and challenges begin to surface. Increasingly visible health and environmental damages are pushing localities to cap coal use. Large power plants with greater minimum outputs are shackling an evolving power grid trying to accommodate new energy sources. Further centralization of ownership is rekindling decade-old political discussions about power sector deregulation and reform.

This unique set of concerns begs the question: how long will coal remain king in China's energy mix?

Managing Conflicts Between Security of Supply and the Environment

China faces the typical conundrum of any large coal-rich nation: as an energy source, coal has an unrivaled security of supply while also being the dirtiest to extract and use. However, whereas other countries enjoy some amount of fuel diversity, China's resource endowments in other fossil fuels are severely limited. China is home to perhaps the world's largest deposits of shale gas, but these resources are still many years away from commercial exploitation.

To meet rising electricity demand, China is rapidly exploiting its renewable and hydropower resources: by 2020, China aims to have 200 GW of wind, 50 GW of solar, 30 GW of biomass, and 300 GW of hydro. Nuclear is expected to increase six-fold, to 80 GW. Yet, even with these massive additions of non-fossil energy sources, China still plans to add 50 GW of coal-fired capacity every year in order to keep up with the projected growth in energy demand.

At the same time, environmental and health damages driven by China's rampant coal use are increasingly apparent. Cancer is the leading cause of death in Beijing, with lung cancer the most common form. Monitoring PM 2.5 is now like checking the weather report — a daily necessity for city — dwellers. A recent study pegged coal's use in northern China to heat homes as contributing to an average loss of 5 years in life expectancy.

In the long term, the International Energy Agency's 450 ppm scenario for climate stabilization calls for an 80% reduction in CO_2 emissions from coal-fired electricity in China by 2035, compared to business as usual, a goal which could only be met with a drastic reorganization of the power sector and significant technological advances in carbon capture. China is currently experimenting with local emission trading schemes to complement administrative energy efficiency measures, which faces a host of challenges before they can really take off.

Coal's Role in China's Economy

Coal is in a semi-regulated state in China. To check inflation, the central government ministry National Development and Reform Commission (NDRC) tightly controls both the retail electricity prices charged to consumers and the electricity tariffs given to coal-fired power plants. At the same time, coal prices are allowed to move more or less with the market rate (NDRC does cap coal contract prices, but it's fairly high and not usually binding). This mismatch between fixed electricity rates and varying coal prices creates significant distortions in the power sector. If coal prices rise, power plant owners and utilities cannot pass along the higher prices to customers by raising electricity tariffs. At irregular intervals, NDRC may allow a generation tariff increase but typically customers do not see the real costs of generation.

In 2011, despite annual electricity demand growth of 12 percent, a steep increase in the price of coal caused many power generators to post record losses. *Last year*, electricity growth dramatically fell to just over 5 percent which, combined with significant new installations of hydropower dams and wind farms, led to a seven percent decline in the average output of China's thermal plants (most of which are coal). Coal prices fell and the major generators made record profits, almost 80% more than in 2011, despite the slow-down in electricity demand growth.

Yet, coal use isn't all about the power sector. Coal directly fuels industrial sectors across the Chinese economy. The iron and steel and manufacturing sectors alone consume as much coal as the entire U.S. fleet of coal-fired power plants. Major coal and power producers are also moving swiftly into coal conversion technologies, such as coal-to-liquids and coal gasification, which ensure a steady demand for the black rock. It was partly due to these ready substitution opportunities across the economy that the coal market was liberalized a decade ago.

Coal Consolidation Policies Target Efficiency

Throughout the coal supply chain, central government policies target consolidation, mostly into the hands of state-owned enterprises. Small mines are being closed down or bought out by large energy groups (100+ million tons annual production is the preferred threshold). The largest coal sector policy in recent years — Small Plant Closure Program — has shuttered old boilers, replacing them with larger and more efficient variants. To shield themselves from price swings, state-owned power producers are increasingly consolidating upstream: Huaneng, one of China's largest electricity generation companies, now reportedly controls 40 billion tons of reserves, roughly ten times China's annual production.

The Small Plant Closure Program, established in the 11th Five-Year Plan (2006-2010), required each locality in the country to identify small power plants, iron and steel smelters, cement kilns and other energy intensive factories to close down and be replaced with larger plants. A state-of-the-art coal-fired power plant capable of reaching higher temperatures and pressures can increase efficiencies by up to 50%. At its peak in 2009, the policy shuttered 26 GW of small coal-fired power plants of 50 MW capacity or less. After closing a group of these old boilers, a single 300 MW or 600 MW plant would be built nearby, with extra capacity to drive new local demand for electricity. This ensured that new commissions always exceed retirements and the total size of the Chinese coal fleet continued to grow. Because of this turnover, China's coal plants are also now more efficient on average than the U.S. coal fleet.

The Consequences of the Policies

The policy was extremely successful largely because it both reinforced an existing technology trend and directly serviced local politicians' investment goals. A single new coal plant together with the energy intensive industrial production it supports leads to a spike in output and looks good on a Chinese official's resume. The replacement program continues today (about 2 GW were closed in the first half of 2012), but the remaining capacity to which it applies is dwindling.

The policy has also led to changes in ownership as many old plants were built and operated by provincial or local governments, whereas new plants are typically centrally-owned. The overriding goal behind unbundling generation and network functions in 2002 was to reduce this monopoly of state control and encourage more market-based functioning, but the largest five power producers still hold on to roughly half the market. Consolidation leads to concentration of political power, which may complicate efforts to achieve reform objectives for the power sector.

Environmental concerns are also driving policies to shift where coal is used. The central government has targeted three regions to institute coal caps during the current 12th Five-Year Plan lasting through 2015. The three regions (centered around Beijing, Shanghai and Guangzhou) accounted for 55% of increased coal consumption in 2011 nationally, mostly from power production. As a first step toward reducing coal consumption, Beijing, Tianjin and Shanghai have all banned or limited new coal plants within their borders, while encouraging plants to use natural gas. When pollution levels rise to dangerous levels, coal boilers, construction and other activities in Beijing are ordered to slow or stop completely.

The effect of eastern coal caps is partly to shift coal-fired electricity production to the west. Massive west-east transmission superhighways are being built by the world's largest grid company, China State Grid, to deliver electricity from the mine's mouth to population centers. Recently, an import ban on low quality thermal coal was floated, which reportedly could reduce imports to the eastern coast by a third, creating more pressure to expand coal-fired power generation in the western mining regions.

Locking in the Future of Coal?

With central government encouragement — as with other areas of the economy — investment in all power generation types has increased. Last year, wind generation growth exceeded the almost flat coal-fired generation growth for the first time ever. This may be a sign of overcapacity: a look at the capacity factors show coal being pushed out while all other sources saw increases — owing partly to a normal precipitation year but also because more nuclear and wind came online. This trend continued into the first half of 2013 with steady thermal power additions while capacity factors declined year-on-year. The latest numbers show capacity currently being built is down from last year, possibly an indication that the central government intends to deal with this overcapacity by slowing new permit approvals.

This coal asset build-up is accelerating fossil infrastructure lock-in, and has huge long-term implications for diversifying China's power mix. The stranded costs of a small, inefficient plant's early retirement may be large but still tolerable. Retiring a 600 MW state-of-the-art plant before its economic lifetime comes at a much steeper price. As a result, the new generation of modern coal plants built in recent years may still be operating in 2050.

In addition, China's large power producers will likely not want their brand new plants sitting idle as alternatives (particularly renewables) capture demand growth. In fact, under current power sector rules, all coal plants are guaranteed "generation quotas" each year to recover costs. After meeting these minima, the cheapest marginal cost (or, most efficient) plant is dispatched. This system in effect sets a floor for coal generation that is hard to get out of without major policy changes.

From the grid standpoint, the larger size of new coal plants increases inflexibility in the operation of the power system. The higher startup costs and minimum generation outputs of larger coal plants makes it more challenging for grid operators to accommodate the variable output of renewable energy sources like wind and solar power. Wind farms in China saw curtailment rates double in 2012. That may mean that while the replacement of aging coal plants improved the efficiency of China's coal fleet and helped reduce both coal use and emissions, it may also present an obstacle to the future growth of low-carbon alternatives to coal.

Dethroning the King?

Coal has shaped the development of the power sector for the last half-century, but with changing public attitudes and new demands from the grid, it may be a two-way street over the coming decades. Health and environmental considerations continue to nudge energy policy, and we may see local coal caps give way to binding regional or national caps that have potential to move the needle. The central government's eagerness to deploy renewables may shift policy surrounding the grid.

Yet, Old King Coal seems to have sure footing when considering the long-term lock-in effects of today's build-out. Overlapping and mixed regulation of the coal sector will complicate designing and implementing effective policies. Barring a rapid boom in shale gas production, coal will continue to offer China unrivaled security and adequacy of supply. China's willingness to face these conflicts may ultimately depend on how prominent its long-term environmental sustainability aspirations are in setting energy sector policy.

Additional Words and Phrases

spurring policy intervention: 激励政策的干预
boiler: 锅炉
undergoing a massive consolidation: 经历一场大规模的巩固合并
unassailable: 不容置疑的,无懈可击的
scratch deeper: 深挖掘
cap coal use: 限制煤炭的使用
shackle an evolving power grid: 束缚发展中的输电网
rekindle decade-old political discussions about: 引发数十年来关于……的政治讨论
power sector deregulation: 电力部门解除管制
conundrum: 难题,谜语
unrivaled security of supply: 至高无上的供给保障
extract: 提取,榨取
resource endowment: 资源储备
shale gas: 页岩气
biomass: 生物量;生物质能
city-dweller: 城市居民
peg coal's use: 长期关注煤炭的使用
ppm: 百万分率
emission: 排放
carbon capture: 碳捕获
complement administrative energy efficiency measure: 补充行政能效措施
take off: 起飞,突然成功
semi-regulated: 半调控的
inflation: 通货膨胀
tariff: 关税

binding：有约束力的
create significant distortions：造成巨大的扭曲
utility：公用事业单位
post record losses：创下亏损记录
wind farm：风力发电厂
swiftly：迅速地，敏捷地
coal conversion technology：煤炭转化技术
gasification：气化
buy out：买下……产权
shutter old boilers：关闭老旧的锅炉
price swing：价格浮动
steel smelter：钢铁冶炼厂
cement kiln：水泥窑
energy intensive factories：能源密集型企业
state-of-the-art：最先进的，最高水准的
commission：委托任务
turnover：营业额
dwindle：缩小；减少
the overriding goal：最重要的目标
institute coal caps：开始实施煤炭管制
float a ban：实行禁令
almost flat：(费率、价格或百分比) 基本固定的
overcapacity：生产能力过剩
a precipitation year：倒退的一年
decline year-on-year：与上年同期数字相比倒退
coal asset build-up：煤炭资产逐步增加或积累
infrastructure：基础设施
the stranded cost：搁置成本
generation quotas：配额生产
meet these minima：达到极小值
dispatch：调度，派遣
inflexibility：不灵活，顽固
startup：启动
variable output：不稳定的产出

curtailment rates：缩减率
nudge energy policy：推动能源政策的制定
deploy renewables：部署可再生能源
build-out：增建，扩建
overlap：重叠
implement effective policies：实施有效的政策
environmental sustainability aspiration：环境可持续发展的愿景

Unit 4 Mining Culture

Foreword

This unit deals with mining culture. After decades of development, miners' life has undergone some changes with the improved environment and more colorful leisure time. In this unit, stories from both Chinese and western miners will lead you to the mysterious mining world, giving an overall scenario of miners' life in different countries, from which the readers can gain insights into both similarities and differences between the Chinese mining culture and western mining culture.

Section One

Preview

Text A gives us an account of a Chinese miner's work and life. Wang Gang, previously a drama college student, followed in his father's footsteps and became a miner. Working in the mine is challenging. Besides the danger and darkness, he suffers physically due to the damp, cold and dusty environment where he works. And life underground is monotonous. But Wang Gang appreciates the miners' eximious qualities and never gives up his hope to better his life. Despite all the hardships, he managed to reach for the outside world and equip himself with skills and abilities needed in case better opportunities come and knock at the door.

Text A

A Miner's Diary

Despite increasing mechanization, life in the cold, dank, dark interiors of a coal mine remains as tough as ever.

Growing up in the rural coal-mining community of Northern Shanxi province, Wang Gang knew just how tough and risky a miner's life can be. And he swore to himself that he would never follow in the footsteps of his father and grandfather.

But destiny had other plans.

Wang, 24, eventually returned to the community he once **spurned**, despite graduating from Shanxi Drama Vocational College in the provincial capital, Taiyuan.

"Given a choice, I would never work in a mine," says Wang dressed in a pair of old jeans, white tennis shoes and a brown artificial leather jacket, dirt **clogging** his long fingernails.

He has traveled a long way from the **halcyon** days of 2002, when the drama college student, a fan of Guns N' Roses, formed a four-member rock band, BY.

After class, the band would do the rounds of the nightclubs of Taiyuan, with Wang its lead singer and guitarist.

But BY did not get progress past being a warm-up band and soon broke up.

Wang then worked as a DJ, and sound and lighting engineer in bars, where he fell in love with a dancer, Zhang Yue.

When Zhang became pregnant, Wang asked her to marry him. Their daughter was born soon after.

He was making a modest 5,000 yuan ($740) a month at that time, which was just enough to get by.

But Wang felt working at nightclubs was full of temptations, and not **conducive** to a stable family life. So, despite the prospects of becoming a partner at the bar where he had worked for years, he quit.

"I like rock music, but I'm also a very conservative man," Wang says.

The couple soon spent all their savings as he could not find another job with a salary decent enough to pay for baby milk and **diapers**.

Wang returned to his hometown, Huairen, a county dotted with dozens of coal mines like countless others in the coal-rich province.

In March 2009, he began life as a miner in Wangping Mine Company, a state-owned mine where his father had once worked.

He clearly remembers how frightened he felt the first time he went down the **maze** of mine tunnels.

"It felt like hell," Wang says.

Typically, a miner works 21 shifts every month to make the maximum he can — 4,000 yuan. Each shift lasts 12 hours and starts at 5 a.m., 1 p.m. and 9 p.m..

Wang works the 1 p.m. shift. After a pre-shift meeting and an **oath** ceremony where the 10-member **squad** of miners vow to ensure safety, he changes into his work **overalls**.

Carrying his **gear** and equipment weighing some 100 kg, he boards a mini-train to descend the mine.

The train comes to a halt half an hour later. He gets off and walks another 20 minutes before reaching his work area.

There is darkness everywhere.

Besides the miners' headlights, there is just one light for every 100 meters along the tunnel.

The ground is cold and moist, with the chill factor **amplified** by **ventilation** fans used to blow fresh air into the mine, and the **sprinkler** that keeps the coal dust down.

Wang has to wear three layers of cotton-padded clothes and trousers to protect himself from the cold and dampness.

In the past, miners used picks and **shovels** to dig the coal and load them onto small carts. Then **mules** would pull the carts to the entrance.

Nowadays, with mechanization, machines are replacing miners to dig the coal.

But Wang, a **newbie**, is unfamiliar with these machines. His main tasks are to assist senior miners, carry equipment and prop up the mine roof with support materials.

His shift ends at 10:30 p.m., by which time his clothes are soaked in sweat and his face covered with black dust.

Taking the same mini-train back to the entrance, he cleans up in the miners' bathroom, and arrives home around midnight.

His wife always stays up until he gets home safe and sound.

All the miners' families are aware of the risks the miners face and live with them. In compensation, they enjoy the highest monthly salaries in the area, which is almost **equivalent** to the annual income of local farmers.

As the world's largest producer and consumer of coal, the country's annual output of coal tripled from 1 billion tons in 1999 to 3 billion tons in 2009, to fuel the fast-growing economy.

Its coal mining industry remains one of the world's deadliest, although the situation has been improving, with the **mortality** rate per million tons declining from 5.71 in 2000 to 0.892 in 2009.

Accidents killed 2,631 coal miners in 2009. That was down from 6,995 deaths in 2002, the most dangerous year on record.

Having worked underground for 18 months, Wang says the worst thing about being a miner is not the danger but the "darkness".

In the winter, he often goes without seeing the sun for weeks, if the shift starts at 5 a.m. and ends at 5 p.m..

"Mechanization and stricter **supervision** have greatly improved mine safety."

"As long as you pay 120 percent attention and follow safety instructions, you will be OK." he says.

Although Wangping, a medium-scale mine with an annual production of 1.5 million tons, has not seen a severe accident in years, Wang says this does not mean no blood has been shed.

Failure to follow instructions can result in **ghastly** wounds such as multiple compound **fractures.**

Wang recalls the utter shock he felt when he chanced upon several fingers of a young co-worker, which were accidentally **severed.**

Wang's experiences have transformed him, body and mind.

While the day-to-day physical work has **bulked** him up, he also suffers from joint pains and a cough, owing to the cold, damp and dusty environment underground.

"It is one of the hardest, dirtiest and most dangerous professions in the world." Wang says.

While miners are often thought of as victims of disasters, he is at pains to point out that "we miners are very brave, **industrious,** are able to work under great pressure, and pay constant attention to detail".

And it is a man's world — state labor laws prohibit the employment of women to work underground.

During breaks and meal times, the hottest topic is women. Who is the best looking girl in the community? Or, who is cheating on whom?

"It is the only entertainment we have underground," Wang says.

The trading of **profanities** and crude insults is common, and helps ease the stress of mine work, he says.

But as a post-80s college graduate, Wang senses a distance between him and older miners.

Of the mine's 6,000 employees, about 1,000 are miners, of which 30 percent belong to the post-80s generation, holding at least a senior high school diploma.

Most of the older miners have just a primary school education and some are **illiterate**. The oldest miner in Wang's team is 44.

"We younger miners have interests the older miners don't have. They don't care who Barack Obama is or what is the breaking news of the moment. They just care about how much money they can earn, or which grocery store has the cheaper eggs."

Few of the educated older miners surf the Internet, while Wang spends almost all his spare time online, teaching himself computer programming.

"Mining has isolated them from the outside world, and made them numb."

Wang says he has to constantly remind himself not to **succumb** to such numbness.

"It is just a job. I won't dig coal forever. I still have dreams," Wang says.

He says he will leave the mine if there are better job opportunities. Older miners tell him he's being impractical, but his family supports him.

For the time being, Wang is focused on a singing contest his company is organizing.

He will sing his favorite song — Chinese singer Wang Feng's *Life in Full Blossom*:

"I have lost my dreams many times;

I want to go beyond the common life;

I want life in full blossom;
Just like flying in the vast sky."

Vocabulary

spurn / spɜːrn / *v.* to reject or refuse somebody or something, especially in a proud way（尤指傲慢地）拒绝

clog / klɒg / *v.* to block something or to become blocked（使）阻塞,堵塞

halcyon / ˈhælsɪən / *adj.* peaceful and happy 平安幸福的

conducive / kənˈdjuːsɪv / *adj.* making it easy, possible or likely for something to happen 使容易（或有可能）发生的；有利于……的

diaper / ˈdaɪ(ə)pər / *n.* a piece of soft cloth or paper that is folded around a baby's bottom and between its legs to absorb and hold its body waste（婴儿）尿布

maze / meɪz / *n.* a system of paths separated by walls or hedges built in a park or garden, that is designed so that it is difficult to find your way through 迷宫；错综复杂；纵横交错；密如蛛网

oath / əʊθ / *n.* a formal promise to do something or a formal statement that something is true 誓言；宣誓；诅咒的话

squad / skwɒd / *n.* a group of people who have a particular task（特殊任务）小组,队

overall *n.* (pl.) trousers that are attached to a piece of cloth which covers your chest and which has straps going over your shoulders〈美〉工装裤；(妇女,小儿等的)罩衣

gear / gɪə(r) / *n.* the equipment or clothing needed for a particular activity（某种活动的）设备,用具,衣服

amplify / ˈæmplɪfaɪ / *v.* to increase something in strength, especially sound 放大,增强（声音等）

ventilation / ˌvent(ə)lˈeɪʃ(ə)n / *n.* the movement or circulation of fresh air 通风,换气

sprinkler / ˈsprɪŋklər / *n.* a device inside a building which automatically sprays out water if there is a rise in temperature because of a fire（建筑物内的）消防喷淋,自动喷水灭火装置

shovel / ˈʃʌv(ə)l / *n.* a tool with a long handle and a broad blade with curved edges, used for moving earth, snow, sand, etc. 铲;铁铲

mule / mju:l / ***n.*** an animal that has a horse and a donkey as parents, used especially for carrying loads 骡子

newbie / ˈnjuːbɪ / ***n.*** a person who is new and has little experience in doing something, especially in using computers novice 新手

equivalent / ɪˈkwɪvələnt / ***adj.*** equal in value, amount, meaning, importance, etc(价值、数量、意义、重要性等)相等的,相同的

mortality / mɒrˈtæləti / ***n.*** the number of deaths in a particular situation or period of time 死亡数量;死亡率

supervision / ˌsupərˈvɪʒ(ə)n / ***n.*** the close watch of people, activities, or places. 监督

ghastly / ˈgæs(t)li / ***adj.*** very frightening and unpleasant, because it involves pain, death, etc.(因有关疼痛、死亡等而)恐怖的,可怕的,令人毛骨悚然的

fracture / ˈfræktʃə(r) / ***n.*** a break in a bone or other hard material(指状态)骨折,断裂,折断,破裂

sever / ˈsevər / ***v.*** to cut something into two pieces; to cut something off something else 切开;割断;切下;割下

bulk ***v.*** to be the most important part of something 使扩大,使形成大量;使显得重要

industrious / ɪnˈdʌstriəs / ***adj.*** working hard; busy 勤奋的;勤劳的;忙碌的

profanity / prəˈfænəti / ***n.*** swear words, or religious words used in a way that shows a lack of respect for God or holy things. 诅咒语

illiterate / ɪˈlɪt(ə)rət / ***adj.*** not knowing how to read or write. 不会读写的;不识字的;文盲的

succumb / səˈkʌm / ***v.*** to not be able to fight an attack, an illness, a temptation, etc. 屈服;屈从;抵挡不住(攻击、疾病、诱惑等)

Phrases and Expressions

do the rounds of: 逐一(条、家)做……
get by: 勉强过活;勉强对付过去
prop up: 支持;支撑
safe and sound: 安然无恙
in compensation: 作为补偿

Unit 4　Mining Culture

chance upon：偶然遇见（碰到，发现）
cheat on：欺骗

Exercises

Ⅰ. Comprehension of the Text

Directions：*Please answer the following questions according to the text.*

1. What was Wang Gang's previous attitude to a miner's job as he grew up in a mine?
2. Why did Wang Gang quit his job in the nightclub?
3. What made Wang Gang become a miner?
4. Can you describe Wang Gang's routine job underground?
5. What negative influence has the job as a miner exerted on Wang Gang's health?
6. What good qualities do miners display, according to Wang Gang?
7. How do miners entertain themselves underground?
8. What are the differences between old miners and young miners?

Ⅱ. Group Discussion

Directions：*Discuss the following questions with your partners, using as much text information as possible in your discussion.*

1. A stable life or an adventurous life, which do you prefer and why?
2. The text tells a lot about the hardships of a miner's life. Can you list some advantages of being a miner?
3. What kind of life is meaningful to you? What do you think of Wang Gang's decision-making when he quitted working in the nightclub and started to work in a mine?

Ⅲ. Word Bank

Directions：*Fill in the blanks with the words on the right side of the text. For each word, you can use only once.*

Due to his eagerness to live a 1. _____ life, Wang Gang, a graduate from Shanxi Drama Vocational College, 2. _____ his job in a nightclub and became a miner in his hometown in order to relieve his economic pressure. However,

a) stable
b) mechanization
c) dusty
d) risky
e) fracture

despite increasing 3. _____, being a miner remains to be as 4. _____ and 5. _____ as ever. The underground world is cold, 6. _____ and 7. _____, owing to which he suffers from joint pains and a cough. The worst of all is he has to face darkness for weeks without seeing the sun. Despite all the challenges, the miners manage to 8. _____ themselves amidst sufferings. Wang Gang feels that his co-workers are brave, 9. _____, and capable of dealing with pressure and details. Wang Gang also finds there are differences between the old miners and the young ones. Money is the only concern of the old miners, whereas young miners try hard to reach the outside world to avoid becoming 10. _____.

f) entertain
g) industrious
h) squad
i) quitted
j) numb
k) newbie
l) supervision
m) tough
n) mortality
o) damp

Ⅳ. Translation Practice

Directions: *Translate the following sentences from Chinese to English.*

1. 尽管政府和个人联合努力,环境问题依然严重。(despite, joint efforts, remain as ... as ever)
2. 只要有耐心,是能成功培育出这一品种的。(given)
3. 除了形成计划,这个项目没有任何进展。(not progress past)
4. 爱丽丝仍然决定去见见她的情敌,那个勤奋却其貌不扬的女人。(rival, industrious)
5. 他十年后才回到故乡,那时双亲已经不在人世了。(not ... until ..., by which time)
6. 学校招收的5,000名新生中,大概1,000名是女生,而其中65%的女生属于管理学院,主修会计和金融。(of ..., of which ...)

Ⅴ. Writing Practice

Directions: *Write a 3-paragraph passage of about 120 words with the title "My Ideal Job" based on the outline below.*

1. Some people dream to be a .... Others harbor the ambition to become a .... My ideal job is ....
2. A perfect plan is formulated to make it true.
3. The prospects of the job are pleasant.

Section Two

> **Preview**
>
> Are you curious about life in a mine? Are you ambitious enough to start a career there? Are you sure that your decision to be a miner a sensible one? Are you optimistic about your job prospects there? What you need is some objective insights into the mine site living. Passage B gives you such a chance by providing both pros and cons of the work and life in a mine. Read it carefully and the text will lead you to the answers as to whether or not mine site living suits you.

Text B

Mine Site Living — What to Expect

Most mining jobs require you to live on-site in a mining camp. Depending on the **roster**, it's not uncommon to spend as long as four weeks **in a row** at the mining village. If you're thinking of getting a job in the mines, you're probably wondering what it's like to live on a mine site. This article discusses what to expect from mine site living: the pros and cons, plus how the right roster can bring success.

Mine Site Living — Pros

Mining companies make a huge effort to keep employees comfortable and happy. By providing good living conditions, companies attract more workers and reduce **turnover** rates. As a result there are some real positives about living in a mining village. These include:

• Sporting Facilities. Employees are encouraged to lead an active lifestyle for both fitness and mental health. Most sites provide a fully equipped **gymnasium** that is free for camp citizens to use at any time of day. Larger mine sites may also **boast** swimming pools, basketball courts, golf driving ranges, football fields, bowling greens and more.

• Meals Cooked for You. There is always a mess hall on-site where employees can eat breakfast, lunch and dinner. The **crossover** of shift workers

(i. e. staff finishing night shifts as others start their day shifts) means all meals are available at all times of day.

• Cleaning Services. Your accommodation is cleaned 2-3 times per week while you're working. When you finish a day's work, you don't need to worry about cooking and cleaning — you're free to relax.

• Pay TV plus Phone and Internet Connection. Nearly all camp sites provide Pay TV plus a phone and Internet connection. This enables you to keep in touch with people back home as well as stay entertained during lonely hours.

• Low Cost of Living (Forced Saving). Although some mine sites have a "wet mess" (i. e. bar or **tavern** facilities), there is little opportunity for shopping. Workers earn excellent money working long days on the mine, but can't waste money on takeaway or dining out. Many people see this "forced saving" as a great advantage in reaching their overall financial goals.

Bear in mind that small mine site may not have all the facilities listed above. Even the smallest of sites, however, will have gymnasium equipment and meals provided.

Mine Site Living — Cons

Mine camp living is not for everyone. The most obvious difficulty is the emotional pressure of being away from friends and family. Here are some more "cons".

• Long Working Days. Most workers living on-site work 7 days a week, 12 hours a day. Working 84 hours a week is exhausting, especially if you're accustomed to the standard 40 hour working week.

• Heat, Dust and Flies. Mines are usually located in remote, hot areas. Dust and flies can become irritating and intense. Summer heat is especially problematic for outdoor workers.

• Camp Food. Food is cooked in bulk portions and usually served in a Bain Marie — or steam tables — in the mess hall. No matter how good the head **chef**, camp food is never quite the same as a home cooked meal.

• Groundhog Day. Workers often complain that mine site living is like the movie *Groundhog Day*, where every day is the same as the last. Most people follow a strict routine. Early morning starts can begin at 4 a. m., followed by long working days and early (7 p. m.) bedtimes. This repetitive lifestyle can be frustrating.

• No Nearby Town. The remote nature of most mining camps means there is no nearby town. There is nowhere to travel to if you want to break the repetitive nature of daily mine-site living.

• Strict Alcohol Monitoring. Shifts often begin with a safety brief or pre-start meeting, which may include alcohol breath testing. Although there might be a bar or tavern on site, workers must limit their alcohol intake in order to be "fit for work" according to health and safety standards.

Roster for Success

Mine site living is vastly improved by finding the perfect roster. People who enjoy living on the mine can more easily tolerate long rosters such as four weeks on, one week off. If you're a parent who needs to spend quality time with your children, you might find an 8 / 8 roster (eight days on, eight days off) is the perfect balance. The industry norm is a two weeks on, one week off roster. Other common rosters are eight days on, six days off and nine days on, five days off.

Aside from your time-on and time-off, you should also consider the day and night shifts you'll be asked to work. A 4 / 4 roster might seem good; that is until you realize you'll be working two days, two nights followed by four days off. Rosters such as this are renowned for disrupting your body clock.

The good news is that mining companies are flexible and will usually offer a roster to suit you—especially if you're a qualified person with skills in high demand.

Is Mine Site Living for You?

How much you enjoy mine site living will depend on your personality and personal circumstances. A single person is more likely to enjoy living in the mine camp than a married person who will miss their spouse. Parents with children usually find the separation even more **traumatic**.

Despite this, working on the mines can bring great career satisfaction. In addition, people often report overall lifestyle improvements as a result of outstanding **remuneration**. If you can strike the right work and life balance with a great salary and good roster, mine site living should be a positive experience.

Vocabulary

on-site / ɒn saɪt / *adj.* 现场的

roster / ˈrɒstər / *n.* a list of people's names and the jobs that they have to do at a particular time 值勤名单；a list of the names of people who are available to do a job, play in a team, etc. 候选名单

pro / prəʊ / *n.* an argument or vote in favor of a proposal or motion 赞成的意见；赞成的理由

con / kɒn / *n.* an argument or vote against a proposal, motion, etc. 反对的理由；反对的投票

turnover / ˈtɜːnəʊvə / *n.* the total amount of goods or services sold by a company during a particular period of time 营业额，成交量；the rate at which employees leave a company and are replaced by other people 人事变更率，人员调整率；the rate at which goods are sold in a shop or store and replaced by others（商店的）货物周转率，销售比率

gymnasium / dʒɪmˈneɪziəm / *n.* a room or hall with equipment for doing physical exercise, for example in a school 健身房，体育馆；physical exercises done in a gym, especially at school（尤指学校的）体育活动

boast / bəʊst / *v.* to have something that is impressive and that you can be proud of 自夸；夸耀

crossover / ˈkrɒsˌəʊvə(r) / *n.* the process or result of changing from one area of activity or style of doing something to another（活动范围或风格的）改变，转型，变化

tavern *n.* a pub or an inn 酒馆；小旅店；客栈

chef / ʃef / *n.* a professional cook, especially the most senior cook in a restaurant, hotel, etc. 厨师；（尤指餐馆、饭店等的）主厨，厨师长

groundhog / ˈɡraʊndˌhɒɡ / *n.* a type of small animal with reddish brown fur that is found in North America 土拨鼠

traumatic / trɒˈmætɪk / *adj.* extremely unpleasant and causing you to feel upset or anxious 痛苦的；极不愉快的

remuneration / rɪˌmjuːnəˈreɪʃən / *n.* an amount of money that is paid to sb for the work they have done 酬金；薪水；报酬

Unit 4 Mining Culture

Phrases and Expressions

in a row：连续，一连
pros and cons：正反两方面；赞成与反对；利与弊；优缺点
bowling greens：草地保龄球

Proper Words and Terms

Bain Marie：a large pan containing hot water in which smaller pans may be set to cook food slowly or to keep food warm 保温汤池；双重蒸锅；加热蒸锅

Groundhog Day：1. (in North America) February 2, when it is said that the groundhog comes out of its hole at the end of winter. If the sun shines and the groundhog sees its shadow, it is said that there will be another six weeks of winter. (北美)土拨鼠日；圣烛节(2月2日，据说是冬末土拨鼠出洞的日子，如果土拨鼠在晴天出洞看到自己的影子，就表示冬天还将持续六个星期)。2. It is a 1993 American fantasy-comedy film directed by Harold Ramis, staring Bill Murray, Andie MacDowell and Chris Elliott. Murray plays Phil Connors, an arrogant Pittsburgh TV weatherman who, during an assignment covering the annual Groundhog Day event in Pennsylvania, finds himself in a time loop, repeating the same day again and again. After indulging in hedonism and committing suicide numerous times, he begins to re-examine his life and priorities. 电影《偷天情缘》，又译《土拨鼠之日》，是一部经典美国喜剧电影，由比尔·莫瑞(Bill Murray)所主演，1993年上映。

Exercises

Comprehension of the Text

Directions：Choose the right answer to each question according to the text.

1. Where do miners live?
    A. a nearby village
    B. a nearby town
    C. a mining camp
    D. a nearby hotel
2. Why do the mining companies make a huge effort to keep employees comfortable and happy?
    A. to charge fees so as to make more money

B. under pressure from the Labor Union

C. under pressure from the government

D. to attract more workers and reduce turnover rates.

3. What does a "Groundhog Day" (paragraph 13) mean?

   A. everyday is the same as the last

   B. a day when spring sets in

   C. February 2

   D. a day when the groundhog emerges

4. How to vastly improve mine site living?

   A. by watching TV

   B. by finding the perfect roster

   C. by having all meals available at all times of day

   D. by leading an active lifestyle for both fitness and mental health

5. Who is / are more likely to enjoy living in the mine?

   A. a single person

   B. a wealthy person

   C. a married person

   D. parents

Section Three

Extended Reading

A Life in the Mine

I work at the Elk Creek Mine, 40 miles from my home in Delta, Colo. I come from a small community where many of the girls in my high school class chose the health care field. That path never appealed to me. Growing up, I was a tomboy and hung out mostly with the guys. I'm athletic, and I played rough. I've never been afraid to get dirt under my nails, so no one who knows me was surprised that I went into mining.

My family has a mining background, so that influenced me, too. My grandfather, brother, stepfather, stepbrother and cousins were all coal miners at one time, and my dad worked in an oil shale mine. In high school, I worked at a restaurant that easily paid my bills, but the company didn't offer health insurance — and mining did. I chose mining.

I've worked here for four years. We work about 2,500 feet beneath the surface. It's about two miles to the deepest part of the mine, and it takes about 30 minutes to get to the work areas.

The first time I entered the mine, I was a little intimidated, because I didn't know what to expect. But it wasn't bad. After that day, I knew I could do the job. There's a lot of shoveling and it's physically demanding, but I enjoy it. I constantly learn new things from my co-workers, who are like a second family.

People have no idea about some of the things we need to know. You take your education underground with you. For example, I use my math skills every day. We have fans that pull fresh air from outside to dilute the dust and methane gas throughout the mine, and we have to keep a close watch on air quantity and quality. Also, we use water for various tasks, like washing the equipment and keeping the coal dust down. We need to know how much water enters the mine so we can make sure it all is pumped out.

During any shift, there are about 60 people in the mine. Five other women work here, but I'm the youngest and the only female on my crew. Some of the men are protective of me and look after me as they would their daughter or granddaughter. From the beginning, I wanted to prove myself and show I could pull my weight.

I started out working on the belt line, the conveyor that ships the coal out of the mine. If the belt malfunctions, the coal can spill, which stops production. So we have to shovel as fast as we can to get it back on the belt. I may not have been able to shovel as fast as the rest of the crew, but I never stopped shoveling when there was a spill. I showed I was capable and willing to do the work.

Last year, I passed the foreman's test, and now I'm a fire boss. I'm one of the first people to enter an area in the mine and check for gas and safety hazards. In the old days, they used flame lamps and a canary, but now we have multi-gas meters that check for several gases at a time.

It's like a little community in the mine. There are underground roads where people drive specially equipped pickup trucks. Some crews can have lunch together in kitchen areas that have microwaves and picnic tables. It's dark, but we have lights on our hard hats and plenty of lights on all the equipment. We also set up lights where we need them. In the fall, we have a company picnic — above ground, of course.

Other than at lunch, there's not much chance for talking. If you can talk and work, OK, but you're always busy. For safety, we wear steel-toed boots and a belt with our name tag and a backup supply of fresh oxygen. You really don't get as dirty as someone might imagine.

You'd be surprised how many people think I'm joking when I tell them I'm an underground coal miner. They usually have three questions. First, they want to know if I mine coal with a pick ax and a shovel. I love this question, because the technology and equipment that we have underground are now so impressive. We do longwall mining, which involves a long, sophisticated steel panel that shears coal off a wall in slices. It's fascinating to watch.

Then people want to know if I'm claustrophobic or afraid of the dark. I tell them I'm not, but a couple of my co-workers have claimed to be. But everyone fears something, and in this job people learn to conquer their fears.

Finally, they ask if it's scary working underground, which to me is practically the same question. I tell them that it's only as scary as you let it be, but that my nature may have a lot to do with my attitude. I'm the type who loves roller coasters. The first time I flew on a plane was for skydiving.

Occasionally, the ground shifts slightly when we work, but it's nothing. It's as if the walls are talking to us. When you think of lightning and monsoons and other events, you realize that Mother Nature can be unpredictable.

When I leave the mine at the end of the day, I have to squint until my eyes adjust to the light. It can be weird to find it's raining or there are two inches of snow after you've been underground all day. We're not allowed to horse around, but if there's fresh snow you almost can't help making a snowball and tossing it at an unsuspecting colleague.

Additional words and phrases

health care field：医疗卫生领域
hang out with：与……一起玩
oil shale mine：油页岩矿
dilute the dust and methane gas：稀释灰尘和瓦斯气体
pump out：泵出
on my crew：在我的团队
pull my weight：履行自己的职责，尽职
the belt line：运输系统
malfunction：故障
spill：溢出；溅出
the foreman's test：班长测试
fireboss：瓦斯监测工
safety hazard：安全危害
flame lamp：火焰灯
canary：金丝雀
multi-gas meter：多种气体检测表
pickup truck：敞篷小型载货卡车
hard hat：安全帽
steel-toed boots：钢制鞋头的靴子
name tag：姓名标签；名牌
backup supply of fresh oxygen：备份的新鲜氧气供给
pick ax：镐斧
steel panel：钢护板
shear：修剪；剪切
slice：片
claustrophobic：幽闭恐惧症的,幽闭恐惧症患者
squint：眯着眼睛看
horse around：胡闹,鬼混

Unit 5 Chinese Coal Mining Industry

Foreword

From the year of 1949 to the present, China's national coal output has been increasing quickly and became first in the world in 2009. Coal mining is playing an important role for the economic development of China. With the improvement of technology and use of mechanized equipment both imported and manufactured locally, the production increases quickly. And at the same time, major coal mining accidents still exist nowadays. In this unit you will get to know a comprehensive knowledge of Chinese coal mining industry from the aspect of coal reserve, production, consumption, development trends, import and export, mining machinery and safety problems.

Section One

Preview

China is rich in coal reserve and coal has always been the most important part in China's primary energy consumption. Coal also contributes greatly to China's GDP growth for the past three decades. It has also been highly self-sufficient. Text A is going to give a brief introduction of the development of China's coal industry from different aspects. From the introduction it's clear to us that coal will remain to be the main energy source for China for a long time and the growth of economy in China will affect the coal industry.

Text A

Brief Introduction to Chinese Coal Mining Industry

China is rich in coal, but poor in other forms of energy resources such as oil and gas. To ensure a steady, reliable energy supply, China relies largely on domestic energy resources to develop its economy. Under this policy, coal has been the dominant component of China's primary energy and accounts for more than 70% of the primary energy consumption, **boldly** carrying the load for sustaining China's 8%-9% annual growth in Gross Domestic Product for the past three decades. The rate of **self-sufficiency** has been above 90%.

Reserves

China has a total proven coal reserve of 997 billion **mt**. They are located mainly in the west and northeast, while population is concentrated in the southeast coastal areas. An overwhelming majority of the reserve are deep and amenable for underground mining only.

Production

China is by far the largest coal producer and consumer in the world. Today, nearly 60% of total production is washed before it is shipped. Since 2003, coal production has been growing at an average annual rate of 250 million mt. At the end of May 2010, production stood at 1.29 billion mt. With 1.1 billion clean tons in 2009, the U.S. is a distant second.

Until the People's Republic of China was established in 1949, coal production was insignificant. From 1949 to 1980, the national strategic mines were established to mine coal under a planned economy in which production, sales and pricing were strictly controlled, and production **fluctuation** was kept to minimum. In the early 1980s, economic reform and a more open policy brought about **vibrant** economic activity in all walks of Chinese life and demands for energy increased sharply. Consequently from 1980 to 1997, in addition to an **accelerated** production increase in national strategic mines, township mines were developed to meet demand. As a result, there were a total of 64,000 mines in 1997 of which 61,000, or 95.3%, were small mines. Under that policy, over-**decentralization** of production caused a supply-demand imbalance and eventually led to a severe **recession** in the coal industry.

From 2000 to present, the coal industry has been reorganized. Coal production is being properly controlled and production distribution is being optimized. Larger coal production groups have been formed and small and medium coal mines are being **consolidated**. Township mines in particular were consolidated with larger national strategic mines under government guidance. Meanwhile, a Chinese-styled market economy was applied to the coal industry.

Coal mines in China are traditionally grouped into three types: national, provincial national and township mines. National strategic mines are mines developed and operated directly by the central government under the original planned economy. In recent years, consolidated small mines have been included. There are a total of 268 national mines located in 22 provinces. In 2008, their production accounted for 50.7% of the overall mine production in China.

Provincial national mines are mines developed and operated directly by the 26 provincial governments in which the mines are located, and in recent years the consolidated small mines have been included. In 2008, their production accounted for 12.7%.

Township mines are those developed after coal markets opened in 1980 and operated by the county and city governments in which the mines are located or by private citizens. They are located in 26 provinces and in 2008, their production accounted for 36.6%.

Strictly, this type of classification of mines was only valid before the Ministry of Coal Industry was abolished in 1997. Thereafter, control and operation of all types of mines was transferred to the provinces. Today only two large groups of mines are controlled directly by the central government, Shenhua and China Coal, while many provincial national mines have been transferred to township and / or private **contractors.**

Consumption

For Chinese coal, there are four major users: electric generation (52%), **metallurgical** (17%), cement (15%), chemical **feedstock** (5%), and others (11%, which includes approximately 6% for residential use). In 2008, Chinese generating capacity stood at 792.53 mw, of which coal accounted for 75.9%, hydro 21.6%, and wind 1%. In 2008, electrical power consumption was 3,433,400 mw/h, of which coal accounted for 81%, hydro 16.4%, nuclear 2% and wind 0.4%.

Steelmakers accounted for roughly 18% of total coal consumption. In terms of total energy used for steelmaking, coal's share, including met, steam and powder coals for **coke** oven, was about 70%.

Chemical feedstock includes the making of fertilizers and other coal-derived chemical products. Notable development in this category was the successful production of oil from coal (coal-to-liquid) at the end of 2008 by Lu'An and Shenhua groups. The plants are capable of producing 5 million mt of oil annually.

Development Trends

Since the implementation of reform and the more open policy in the early 1980s, coal supply has advanced from shortage to a near **equilibrium** state. The coal economy has evolved from planned to Chinese-styled market economy.

Production has changed from small-scale numerous coal mines to **conglomerates** of extra-large mines. For example in 2003, the numbers of mines were reduced from more than 81,000 to about 25,000. That figure was further reduced to 15,000 in 2009. Development of coal mining conglomerates has begun, the first of which went into production in December 2008. Under this policy, certain trends occur.

1) Mechanization improved considerably in recent years, more than 90%, in national strategic mines with productivity reaching 5.06 mt / man.

2) Coal mining companies have expanded from single product enterprise, i.e., producing coal, to multiple coal-based enterprises to include electric generation, aluminum, chemical, construction, coking and coalbed methane.

3) Fatalities for all coal mines have decreased steadily from 9.44 per million raw tons in 1978 to 1.485 in 2007.

4) Continue to **renovate** medium- and small-sized coal mines and shut down small ones not conforming to industrial policies, with poor safety conditions, wasting resources and harming the environment, so as to further **optimize** the structure of the coal industry.

5) Promote the coordinated development of related industries, and encourage coal-electricity joint operation or coal-electricity-transport integrated management, so as to extend the coal industry chain.

Coal is and will be the main energy source for China and it will remain that way for a long time to come. This is not to say coal supply and demand will not experience supply shortage or oversupply in any types of coals or in some districts or within some brief periods. Coal consumption and consequently production in China will grow in response to national economic growth in the long run.

Vocabulary

boldly / ˈbəuldlɪ / ***adv.*** showing no fear 大胆地,冒失地
sufficiency / səˈfɪʃ(ə)nsɪ / ***n.*** enough quantity 足量,充足
reserve / rɪˈzɜːv / ***n.*** something that is being or has been stored for later use 储备,储存
mt：metric ton 公吨
fluctuation / flʌktʃʊˈeɪʃ(ə)n / ***n.*** moving up and down 起伏,波动
vibrant / vaɪbrənt / ***adj.*** thrilling, vibrating 振动的,充满生气的
accelerate / əkˈseləreɪt / ***v.*** increase the speed of, cause to move faster or happen earlier 使……加快,使增速
decentralization / diːˌsentrəlaɪˈzeɪʃən / ***n.*** giving greater powers (for self-government, etc.) to (places, branches, etc. away from the center) 分散,非集权化
recession / rɪˈseʃ(ə)n / ***n.*** withdrawal, act of receding 衰退,不景气
consolidate / kənˈsɒlɪdeɪt / ***v.*** make or become solid or strong 巩固,联合
contractor / kənˈtræktə / ***n.*** person, business firm, that enters into contracts 承包人
metallurgical / metəˈlɜːdʒɪkl / ***adj.*** of or relating to metallurgy 冶金的,冶金学的
feedstock / ˈfiːdstɒk / ***n.*** the main raw material used in the manufacture of a product 原料
coke / kəʊk / ***n.*** light substance that remains when gas has been taken out of coal by heating it in an oven, used as fuel in stoves and furnaces 焦炭,焦煤
equilibrium / iːkwɪˈlɪbrɪəm / ***n.*** state of being balance 平衡,平静
conglomerate / kənˈglɒm(ə)rət / ***n.*** a large company formed by joining together different firms 企业集团
renovate / ˈrenəveɪt / ***v.*** restore old buildings, oil paintings, etc. to strong condition 更新,修复,革新
optimize / ˈɒptɪmaɪz / ***v.*** to arrange or design something so that it operates as smoothly and efficiently as possible 使最优化,使完善

Unit 5　Chinese Coal Mining Industry

Phrases and Expressions

be located in：位于，坐落于
an overwhelming majority of：绝大多数
supply-demand imbalance：供求失衡
under government guidance：在政府的指导下
Chinese-styled market economy：中国特色的市场经济

Exercises

Ⅰ. Comprehension of the Text

1. How many percentages does coal account for China's primary energy consumption?
2. According to the text, is it true that the majority of coal reserve in China is for underground mining only?
3. Which country is the second one in coal production in the world according to the text?
4. What was the cause of severe recession in the coal industry in the year of 1997?
5. What happened to the coal industry after 2000?
6. How many traditional types of coal mines are there in China? What are they?
7. When was the Ministry of Coal Industry abolished? What changes happened after that?
8. Can oil be derived from coal according to the text? Cite one example if you can.

Ⅱ. Group Discussion

1. How do you understand the small mines owned by private contractors? What kind of roles it has played in China's coal industry development?
2. Talk to your partners about the development of coal industry in China from 1949. Summarize it in your own words.
3. Do you know any chemical products derived from coal besides oil? Talk to your partners and try to find more about them.

Ⅲ. Word Bank

Directions: *Fill the blanks with the words on the right side of the text. For each word, you can use only once.*

China is by far the largest coal producer in the world. The most 1. _____ are for underground mining only. The economic reform brought about 2. _____ economic activities and 3. _____ coal production in China. Coal supply advanced from shortage to 4. _____ state. But the 5. _____ of coal production caused a supply-demand imbalance in China, since there were more than 61,000 small mines owned by private 6. _____ which eventually led to a severe 7. _____ in the coal industry. After 2000, many small mines were 8. _____, medium-sized mines were 9. _____ so as to further 10. _____ the structure of the coal industry.

a) equilibrium
b) vibrant
c) decentralization
d) coke
e) development
f) reserve
g) accelerated
h) economy
i) contractors
j) steelmakers
k) optimize
l) shut down
m) industry
n) renovated
o) recession

Ⅳ. Translation Practice

Directions: Translate the following sentences from Chinese to English.

1. 中国人口主要聚集在东南沿海一带。(concentrate)
2. 经济改革和对外开放政策给中国的各行各业带来了充满活力的经济活动。(all walks of life)
3. 供需不平衡导致煤炭行业的严重衰退。(supply-demand imbalance)
4. 2000年之后,中小型煤矿被加以整合,形成更大的产煤企业(consolidate)
5. 2008年,国有大型煤矿的产煤量占全国总产煤量的50.7%。(account for)
6. 长久来看,中国的煤炭消费和产出将与中国的经济增长相呼应。(in the long run)

Ⅴ. Writing Practice

Directions: Write a 3-paragraph passage of about 120 words with the title "On the Advantages and Disadvantages of Small Mines in China" based on the outline below.

1. Small mines ever played important roles in China's economic development.
2. Small mines also cause a lot of devastating accidents in mines around China.
3. It is crucial to reregulate and even shut down some small mines.

Section Two

Preview

Coal mining accidents have always been a major problem in the history of China's coal mine development. Text B introduces the major safety problems occurring in Chinese coal mines. Chinese government took great efforts to reduce fatalities in coal mines. Some township small mines were shut down, for they contributed a larger part of fatalities in Chinese coal mines. The government also consolidates some small mines and put them under the supervision of national strategic mines. Aside from safety issue, coal mine machinery, coal import and export situation are also introduced.

Text B

Safety

In the process of changing administrative control of national industries in 1997, the central government decided to abolish the ministries of several industries including coal. The State Administration of Coal Mine Safety (SACMS) was organized to manage and deal with the severe mine accident issues in China. It has 26 provincial branch offices in coal-producing provinces with a total of 73 district offices.

Similar to Chinese mining technology, the accident rates vary widely for different mining groups. They have many safe mines but they also have many mines employing unsafe mining practices. Accident rates such as the fatality rate is defined by number of fatalities per 1 million mt raw of coal production. The fatality rate has been decreasing steadily for the past 30 years. For national strategic mines, the fatality rate in 2008 was 0.33. Aside from the direct causes discussed below, the imbalance between areas of production and demand provides the opportunity for the unsafe small township mines to continue to contribute to the fatality numbers.

Major Types of Accidents

Township mines were the major problem areas. They produced 37% of the total coal but accounted for 73% of the fatalities.

The major categories of accidents consisted of gas, flooding, roof fall, fire and **haulage**. Gas accidents included primarily coal and gas outbursts, followed by methane explosions. They occurred more at the faces of roadway development and in the township mines. The causes for methane explosions included either no methane monitoring system or if it existed, improperly installed or insufficient sensors, improper local **ventilation** or mine ventilation system, and illegal **blasting** and electric equipment. Coal and gas outbursts occurred mostly in geologically disturbed areas or where coal seams were subjected to rapid geological changes.

Flooding accidents occur the most at the roadway development faces cutting into the abandoned mines that were flooded with water. The major causes were unknown geological conditions, faulty advance water detection procedures, or ignorance of legal requirements.

Roof fall accidents occurred largely in township mines, and in descending order of frequency, at the coal mining faces, roadway development faces, roadways, **shafts** and **gob**. The major causes for roof falls were no support and poor quality of supports.

Mine fire accidents were due to spontaneous combustion of coal in the gob areas that were not sealed properly or timely, mismanagement, no monitoring system for hazardous gases, and poor ventilation systems.

Haulage accidents involved mainly **hoisting** equipment due to poor maintenance, brake failures, wire rope failure, mine car collisions, etc.

Methods for Fatality Reduction

The governments have been working very hard to reduce the accident or fatality for quite a long time. In addition to strengthening the enforcement of safety supervision of coal mines, and guiding local governments and enterprises to intensify efforts in technological upgrading for coal mine safety, the construction of safety facilities, and shutting down the illegally operated small township mines, the most effective measure is to close small township mines.

In 2007, 11,155 township mines were closed. More than 1,000 township mines were closed in 2008 and 2009 respectively, eliminating approximately 250 million tons of production capacity. The target is to close small mines to less than 10,000 in 2010. The **on-going** approach to deal with those township mines is that small township mines surrounding the national strategic mines are being consolidated and put under the supervision of the corresponding national strategic mines. Under this policy, small township mines began to employ proper mining methods and safe practices. Consequently the fatality rate for township mines has been reduced considerably from 5.5 in 2005 to 2.337 in 2008.

Coal Mining Machinery

In 2007, there were a total of 115 Chinese manufacturers making all types of coal mining machines and the number of manufacturers increased on average five per year. For instance, there are more than 160 models of Chinese shields. As many as 14 Chinese manufacturers make 546 shearers (509 double-drum ranging arm, 35 single-drum ranging arm and three plows) in 2007. A total of 32 companies build 725 AFCs (Armored Face Conveyor).

China also imports equipment. Imported mining machines can be grouped into two types: complete and partial sets. In the complete set category, the whole set of face equipment was imported, mainly because no comparable quality of products existed in the domestic market. The best example is the Shenhua group in Inner Mongolia where the geological conditions are similar to U.S. For the past 15 years they had imported many sets of longwall face equipment from the U.S. and Germany. In the partial set category, one or more types of face machine were imported while the remainders were from domestic manufacturers. The typical example was the Tiefa Group's plow system in which the plow, AFC and shield's electro-hydraulic control system were imported from Germany while shields were made domestically.

Import and Export of Coal

Chinese coal exports are controlled and a **quota** is distributed by the central government to a few coal groups. It is adjusted periodically depending on the domestic situation of supply and demand. Major receiving countries and areas for steam coal exports were India, Japan, South Korea and Taiwan, while Brazil, India, Japan and the U.S. imported Chinese coke.

More recently, however, China has become a net importer of coal. And, the import policy for coal is more liberal. The import climate is most **ripe** for the southeastern coastal areas where it is farthest from the coal producing areas, while the demand is the greatest due to concentrated population. Fujian, Guangdong and Guangxi are the three major provinces importing coals. Preliminary statistics shows for the first five months of 2010 net import of coal totaled 60.11 million tons.

Vocabulary

haulage [ˈhɔːlɪdʒ] *n.* transport of goods 托运,拖拽

ventilation [ˌventɪˈleɪʃən] *n.* 通风设备

blasting [ˈblɑːstɪŋ] *n.* the strong rush of air or gas spreading outwards from an explosion 爆炸,破坏

shaft [ʃɑːft] *n.* long, slender stem of an arrow or spear 机轴

gob [ɡɒb] *n.* 块状物,采空区

hoisting [ˈhɔɪstɪŋ] *v.* lift with an apparatus of ropes or a kind of elevators 提升,举重

on-going [ˈɒnɡəʊɪŋ] *adj.* 不间断的,进行的

quota [ˈkwəʊtə] *n.* limited share, amount or number, esp. a quantity of goods allowed to be manufactured, sold, etc. 配额,定额

ripe [raɪp] *adj.* fully developed 成熟的,时机成熟的

Exercises

Directions: Choose the right answer to each question according to the text.

1. Which type of mines contributes to the most fatalities?
   A. national strategic mines
   B. township mines
   C. medium sized mines
   D. international cooperation

2. Which of the following is not included as the major safety accident according to the text?
   A. gas
   B. flooding
   C. roof fall
   D. management

3. Which of the following is not the major cause for flooding accident?

A. unknown geological condition
B. faulty advance water detection procedure
C. ignorance of legal requirement
D. rain

4. Which country does not import Chinese coke?
   A. Brazil
   B. India
   C. U. S.
   D. South Korea

5. Which province does not belong to the major import coal province?
   A. Fujian
   B. Guangdong
   C. Guangxi
   D. Sichuan

Section Three

Extended Reading

Introduction of Coal—A Complex Natural Resource

Stanley P. Schweinfurth

Coal is abundant in the U. S. coal is relatively inexpensive, and is an excellent source of energy and byproduct raw materials. Because of these factors, domestic coal is the primary source of fuel for electric power plants in the U. S., and will continue to be well into the 21st century. In addition, other U. S. industries continue to use coal for fuel and coke production and there is a large overseas market for highquality American coal.

Because humans have used coal for centuries, much is known about it. The usefulness of coal as a heat source and the myriad of byproducts that can be produced from coal are well understood. The continued and increasingly large-scale use of coal in the United States and in many other industrialized and developing nations has resulted in known and anticipated hazards to environmental quality and human health. As a result, there is still much to be learned about the harmful attributes of coal and how they may be removed, modified, or avoided to make coal use less harmful to humans and nature. These issues of coal quality have not been examined carefully until recently.

Some of the problems that accompany the mining and use of coal are well known. Acid mine drainage results when coal beds and surrounding strata containing medium to high amounts of sulfur, in the form of compounds known as sulfides, are disrupted by mining, thereby exposing the sulfides to air and water. Atmospheric sulfur oxides (SO_x) and subsequent acid deposition (such as acid rain) result from the burning of medium- to high-sulfur coal. The quality of surface and ground water may be affected adversely by the disposal of the ash and sludge that result from the burning of coal and cleaning of flue gases. These are some of the serious problems requiring either improved or new remedies. Other environmental problems are associated with emissions of carbon dioxide (CO_2) and nitrogen oxides (NO_x), two of the so-called "greenhouse gases." These emissions are often attributed to coal use only; however, they also result from the burning of any fossil or biomass fuel, such as wood, natural gas, gasoline, and heating oil. The greenhouse gas problem requires a broader solution than just reducing the use of coal. Research currently is being conducted in the U.S. and several other countries into the reduction and disposal of CO_2 from coal combustion. An excellent review of the results of this research and the prospects for coal can be found in a publication of IEA Coal Research (1999).

Fluidized-bed combustion (FBC) of coal, which is a new method for controlling sulfur emissions, is gaining wide acceptance. In this system, finely ground coal is mixed with finely ground limestone. Both are fed together into a furnace in a constant stream onto a horizontally moving grate. Air is forced up through the grate and the entire mass is ignited at relatively low temperatures. The forced air causes the ground coal and limestone to be mixed with the hot gases of combustion, which in turn promotes the conversion of any SO_x to gypsum as the burning mass moves along on the grate. According to the U.S. Department of Energy (2000), high-sulfur coal may be burned in this way while simultaneously capturing up to 95 percent of the SO_x and most of the NO_x emitted.

Reduction and disposal of CO_2 emissions is the subject of current research in the U. S. and several other countries.

The Clean Air Act Amendments of 1990 (Public Law 101-549) required the U. S. Environmental Protection Agency (EPA) to conduct studies of 15 trace elements released by the burning of coal to determine if they present health hazards. These 15 elements (antimony, arsenic, beryllium, cadmium, chlorine, chromium, cobalt, lead, manganese, mercury, nickel, potassium, selenium, thorium, and uranium), along with many other potentially hazardous substances released into the air by other industries, are termed "hazardous air pollutants" (HAPs).

On the basis of epidemiological studies, the EPA concluded that, with the possible exception of mercury, there is no compelling evidence to indicate that trace-element emissions from coal-burning power plants cause human health problems (U. S. Environmental Protection Agency, 1996). In December 2000, after extensive study, the EPA announced a finding that regulation of mercury emissions from coal-fired power plants is necessary and appropriate because coal-fired power plants are the largest unregulated source of manmade emissions of mercury in the U. S. (U. S. Environmental Protection Agency, 2000).

EPA will propose regulations to limit mercury emissions in 2003 and will issue final emissions rules by 2004. Meanwhile, arsenic is still under study, not as an emissions problem from coal combustion, but for its potential role as a hazard in ground water if it is leached from coal-mining waste or from fly ash in disposal sites. Additional coal-quality research on both mercury and arsenic is being conducted at the U. S. Geological Survey (USGS) and elsewhere in order to (1) help identify their sources in coal, and (2) help resolve any remaining potential hazards issues with regard to these two elements.

In 1996, the EPA said that, of the potential HAPs from coal combustion, only mercury and arsenic needed additional study. In 2000, EPA determined that mercury emissions require regulation.

Other problems that may arise from the use of coal are not necessarily harmful to the environment or human health, but affect the use and efficiency of coal-burning equipment. For example, certain constituents in coal may cause severe erosion and corrosion of, or the buildup of mineral deposits on, furnace and boiler parts. These effects greatly reduce the efficiency and life expectancy of furnaces or boilers; as a result, costly repairs are often required.

Thus far, coal has been discussed as if it were a single homogeneous material, but this is done only for convenience in this report. In fact, a wide variety of coal types exists, depending on differing proportions and combinations of organic and inorganic (mineral) components, sulfur, and ash. For example, sulfur content may range from low (less than 1 percent), through medium (1 to 3 percent), to high (greater than 3 percent), and ash yields may range from a low of about 3 percent to a high of 49 percent (if ash yields are 50 percent or greater, the substance is no longer called coal). Coal may produce high or low amounts of energy, or contain high or low amounts of the substances that produce organic chemicals and synthetic fuels, or contain higher or lower amounts of the elements that are considered to be hazardous air pollutants (HAPs). This range in properties results from coal's diverse origins, including the long and complex geologic histories of coal deposits.

(*The passage is an extraction from the report* Coal—A Complex Natural Resource *and adapted by the editors of this course book. The original report can be retrieved at https://pubs.usgs.gov/circ/c1143/c1143.pdf.*)

UINT 6　Five Major Mining Disasters

Foreword

In this unit, you will learn about the five major Mining Disasters. It is commonly known that coal fire, methane, dust, coal mine water and roof fall are the greatest threats to the minors' safety, environmental problems, and valuable natural resources. It will also cause huge financial losses to mines both in China and around the world. It is vitally important to understand the formation of the five mining disasters and adopt proper methods to get rid of them before they occur. Formation theory and controlling technology of each disaster are introduced respectively in this unit.

Section One

Preview

Text A gives an overall introduction to the five major mining disasters. Coal fires devour enormous coal resources and emit large amounts of greenhouse gases. The explosion risk of methane is a serious threat to mining safety. Serious lung hazard to worker is often caused by prolonged exposure to excessive levels of coal dust. Karst water inrush will worsen or even damage the mining sites and thus lead to flooded coal mine and death of miners. The most dangerous threat to mining and miners is roof fall. When roof fall occurs most workers suffer from injury, disability, let alone many cases of fatality.

Text A

Five Major Mining Disasters

1. Coal Fire

Coal fires occur most frequently in exposed or underground coal seams.

They are often triggered via spontaneous **combustion** of coal, which is an exothermic oxidation reaction between coal and air. Coal temperature will increase if heat released by coal oxidation is not sufficiently **dissipated**. When the temperature reaches a critical temperature (80 ℃ — 130 ℃), the coal will start to smolder and burn. A large number of factors influence spontaneous combustion of coal, which can be categorized into three groups: coal properties, coal seam properties and external conditions. Apart from factors influencing the spontaneous combustion of coal, forest fires, lightning, and human activities such as mining activities, **discarding** of burning cigarettes, burning trash piles, and ignition of coal for heating, cooking, and generating "Red Dog" for ash used for traction on ice also trigger coal fires. Among these factors, mining activities exposing coal to air or creating **ventilation** pathways are the most common factors triggering coal fires.

Coal fires pose great threats to valuable energy resources, the environment, human health, and safety. They devour enormous coal resources and emit large amounts of greenhouse gases. Toxic gases released by coal fires threaten local habitant's health. Unstable overlying rocks, large cracks, and **subsidence** induced by coal fires are hazardous for both **infrastructure** and people.

China, as the largest coal producer in the world, has suffered from the serious hazards of coal fires since the 1960s. Many international and Chinese experts and scholars have investigated coal fires in China.

2. Methane

Coal mine methane (CMM) is a general term for all methane released mainly during and after mining operations.

Coal mine methane has always been considered as a danger for underground coal mining as it can create a serious threat to mining safety and productivity due to its explosion risk. One of the most important duties of ventilation in underground coal mines is to keep methane levels well below the explosive limit by **diluting** methane emissions that occur during mining.

In addition to proper ventilation practices, removal of coal mine methane from the mining environment prior to, during, and after coal production by using various in-seam and surface-to-mine borehole designs, has been the key component to **alleviate** the explosion threat in mining operations.

Gas content is one of the key data included in coalbed methane resource estimations. During mining, all three components of coal gas content (lost, desorbed, and residual) can potentially contribute to methane emissions into the mine atmosphere.

3. Dust

Fugitive dust on longwalls has always been an issue of concern for production, safety and the health of workers in the underground coal mining industry globally. **Respirable** dust **particles** have long been known to be a serious health hazard to workers. Prolonged exposure to excessive levels of airborne respirable coal dust can lead to coal workers' lung problems such as **pneumoconiosis**, which is **irreversible** and can be **debilitating**, progressive, and potentially fatal in their most extreme cases.

Dust particles can be generated from several sources on the longwall, primarily including shear cutting, chock movements, stage loader / crusher (BSL) and intake contaminations. Studies indicate that longwall shearer and chocks are the main dust sources on longwall faces, representing up to 80% of the total dust produced. The drive for high production output has led to the advancements in longwall technologies of more powerful and faster shearers, and consequently the requirement for chocks to move at a faster rate. As the supports are lowered and advanced, crushed coal and / or roof rock drops from the top of the **canopy** into the air flow on the longwall face. As a result, chock movements can be a significant source of respirable dust for shearer operators when chocks are advanced upwind of the shearer during a main gate to **tailgate** cut.

4. Coal Mine Water

As **hydrogeological** conditions of coal field in North China are complicated, coal mine water hazards have been occurring frequently.

Coal field of North China is located in Shanxi, Hebei, Henan, Shandong, Jiangsu and Anhui province. Its area is about 727,600 km^2. Its coal production is about 60% of the whole country, so it has an important role in China energy system. But there is a problem of coal mine water hazards in North China, 80% of coal mines are affected by Karst water. According to rough statistics, in recent 50 years, 200 incidents of water inrush have occurred, 1,500 persons have died, and there is an economic loss of 3 billion

Yuan (RMB).

Karst water inrush has following properties: groundwater **yield** is huge and steady. Once water inrush occurred, the working condition would become worse. More seriously, working face, mining area and coal mine would be flooded and thus many persons would die. Besides, the water hazards would cause the replacement of mining area and level stretch very difficult and influence the production of coal mines. Water inrush would also increase coal mine water drainage. The cost of electricity to drain water would increase. For example, in Jiaozuo coal field (Henan province), 62 cubic meters mine water was drained in the process of excavating 1 ton of coal. Stronger coal mine water drainage would bring a series of environmental problems, such as water quality worsen, water resource exhausted, surface collapse emerged and mine water contaminated. For example, in 1954, Jinci Spring (Shanxi province) yield was 2.0 cubic meters per minute, when Xishan coal field was exploited, the yield was dropping. Until May 1994, Jinci Spring exhausted.

5. Roof Fall

There are several reasons that make underground coal mining one of the most hazardous activities, and the most important one is roof fall. Roof fall is the greatest safety hazard that underground coal miners deal with. Roof fall can cause **detrimental** effects on workers in the form of injury, disability or fatality and also on mining companies because of downtimes, interruptions in the mining operations, equipment breakdowns, etc. The hazardous nature of roof fall can be illustrated by the statistics of mine accidents. For example, US mine accident statistics indicated that during 10 years, 1996 — 2005, 7,738 miners were injured from roof falls in underground coal, metal, nonmetal and stone mines. Coal mines showed the highest injury rate, 1.75 injuries per 200,000 h underground work. Fatal injury trends from 1996 to 2005 were equally troubling, with 100 roof fall fatalities, while coal mines had the highest number of 82 (0.021 fatalities per 100,000 miners). In 1998, a total of 2,232 unplanned roof falls occurred in 884 US underground coal mines. These falls resulted in 419 injuries and 13 fatalities.

Unplanned roof failures in coal mines can be created by a number of different factors. These include geologic defects in the roof rock, moisture degradation of **shale**, extreme loading conditions under high cover, multiple seam mining and inadequate support, etc.

Vocabulary

combustion / kəmˈbʌstʃən / ***n.*** an act of burning something or the process of burning 燃烧, 氧化

dissipate / ˈdɪsɪpeɪt / ***v.*** things become less and less strong until it goes away completely 浪费, 使消散

discard / dɪˈskɑːd / ***v.*** to get rid of something 放弃, 丢弃

ventilation / ventɪˈleɪʃ(ə)n / ***n.*** an act of letting fresh air to get into 通风, 通风设备

subsidence / ˈsʌbsɪd(ə)ns / ***n.*** going down of the level of water, especially flood water 下沉, 沉淀

infrastructure / ˈɪnfrəstrʌktʃə / ***n.*** basic facilities such as transportation, communication, power supplies and buildings, which enable it to function properly 基础设施, 公共设施

dilute / daɪˈl(j)uːt / ***v.*** to make a liquid weaker by adding water or another liquid 稀释, 冲淡

alleviate / əˈliːvɪeɪt / ***v.*** to ease the intensity of suffering, pain or an unpleasant condition 减轻, 缓和

respirable / rɪˈspaɪəbl / ***adj.*** able to be breathed 可呼吸的, 能呼吸的

particle / ˈpɑːtɪkl / ***n.*** very small pieces or amount of something 颗粒, 质点

pneumoconiosis / ˌnjuːməkəʊnɪˈəʊsɪs / ***n.*** any disease of the lungs or bronchi caused by the inhalation of metallic or mineral particles, characterized by inflammation, cough, and fibrosis 尘肺病

irreversible / ɪrɪˈvɜːsɪb(ə)l / ***adj.*** that cannot be reversed or revoked 不可逆的, 不能翻转的

debilitate / dɪˈbɪlɪteɪt / ***v.*** make weak 使衰弱, 使虚弱

canopy / ˈkænəpɪ / ***n.*** (usu. cloth) covering above a bed, throne, etc. 天篷, 遮篷

tailgate / ˈteɪlgeɪt / ***n.*** door or flap at the rear of a motor-vehicle which can be opened for loading and unloading 后挡板, 下闸门

hydrogeological / ˈhaɪdrəʊ-dʒɪəˈlɒdʒɪkl / ***adj.*** 水文地质的

yield / jiːld / ***n.*** amount produced 产量, 受益

detrimental / ˌdetrɪˈment(ə)l / ***adj.*** harmful 不利的,有害的

shale / ʃeɪl / ***n.*** a sedimentary rock formed by the deposition of successive layers of clay 页岩,泥板岩

Exercises

Ⅰ. Comprehension of the text

Directions: Please answer the following questions according to the text.

1. Where does coal fire most frequently occur?
2. What's the critical temperature for the coal to smolder and burn?
3. Why is coal mine methane a risk to mining production?
4. What's the key component to alleviate the explosion threat in mining operations?
5. What's the main cause that leads to coal workers' pneumoconiosis?
6. Which provinces in North China suffer greatly from coal mine water?
7. What are the serious consequences caused by coal mine water?
8. What are the factors that cause roof fall in underground mining?

Ⅱ. Group Discussion

Directions: Discuss the following questions with your partners, using as much text information as possible in your discussion.

1. How do you think human activities can trigger coal fire? Talk to your partners and name some of them.
2. Can you explain the procedures of the advancements in longwall faces in simple language?
3. What are the environmental problems caused by coal mine water drainage? Cite one example in the text.

Ⅲ. Word Bank

Directions: Fill the blanks with the words on the right side of the text. For each word, you can use only once.

Coal fires are often triggered through 1. _____ combustion of coal. It not only poses great threats to energy resources, but also 2. _____ large amounts of greenhouse gases. Unstable overlying rocks are dangerous for both 3.

a) infrastructure
b) drainage
c) yield
d) respirable
e) spontaneous

_____ and people. One of the important component to 4. _____ the explosion threat is to remove the coal mine methane from the mining environment, for it creates a serious threat to mining safety. 5. _____ dust 6. _____ are serious health hazard to workers. Pneumoconiosis caused by coal dust is 7. _____ on the longwall face. Huge groundwater 8. _____ is one of the properties of Karst water inrush. Strong coal mine water 9. _____ would bring serious environment problems. Besides coal mine water, roof fall poses the most 10. _____ mining activities which lead to fatal injuries to mine workers.

f) emit
g) irreversible
h) particle
i) hazardous
j) alleviate
k) geological
l) multiple
m) drained
n) exhausted
o) complicated

Ⅳ. Translation Practice

Directions: Translate the following sentences from Chinese to English.
1. 煤矿火灾对宝贵的能源资源构成巨大的威胁。(pose ... threats to)
2. 自上个世纪 60 年代起,中国就深受严重的煤矿火灾之苦。(suffer from)
3. 由于有爆炸的危险,煤层气一直被认为是井下采煤的危害之一。(due to)
4. 长期地暴露在大量的煤尘中会使煤矿工人患上严重的肺病。(expose to)
5. 长壁工作面的煤尘由多种来源产生。(be generated from)
6. 中国北方的主要产煤区位于山西、陕西、河南、内蒙古等省份。(be located in)

Ⅴ. Writing Practice

Directions: Write a 2-paragraph passage of about 120 words with the title "Coal Mine Water" according to the outline below.
1. Talk about the formation of underground coal mine water.
2. What are the hazardous consequences of coal mine water?

Section Two

Preview

 This section gives an overall introduction to controlling technologies of each disaster. High technology such as space-borne remote sensing investigation is applied in detecting and monitoring of coal fires. Experts adopt the separation technology of gas mixture to concentrate methane so that it can be used as fuel for gas engines, boilers, etc. Proper use of water is still the most effective way of dust control in mines. The concept of green mining is put forward to preserve environment and maximize the economic and social benefits; coal water is also included to be protected while mining. Sufficient awareness of effective parameters is commonly adopted in roof fall prevention.

Text B

Updated Technologies in Controlling of the Five Major Mining Disasters

1. Detection and Monitoring of Coal Fires

Over the last decade, techniques to detect and monitor coal fires have developed quickly and numerous advanced findings were published. With respect to the level of **spatial** detail and detection accuracy desired, coal fire detection and monitoring can be categorized into four groups: underground-, ground-, airborne-, and space-borne investigations.

1) Underground detecting techniques

Borehole drilling is an important underground detecting technique. However, during the last decade, underground detecting techniques mainly focused on index gas and radon gas detections. The index gas method aims to search for a relationship between emitted gases and the temperature of a specific coal. Consequently, based on laboratory scale experiments with regard to the self-heating and ignition of coal, coal temperature can be predicted according to collected gases, which is significant for early warning of self-ignition.

Radon gas measurement is a promising method to detect coal fires, especially for newly ignited areas. It has been widely used to detect the underground **spontaneous** combustion of coal in China.

2) Ground detecting techniques

Ground detecting techniques include the following: Temperature measurement, self-potential method, 2-D electrical imaging and electromagnetic techniques, magnetic techniques and ground penetrating radar.

3) Airborne remote sensing

Airborne remote sensing has the advantage of flexible data acquisition, flexible flight height and high spatial resolution, and the opportunity to cover a large area. During the last decade, research on airborne remote sensing for coal fire detection in China mainly focused on airborne **thermal** remote sensing, the use of unmanned aerial vehicle (UAV) with a thermal camera, and hyper spectral (the Operative Modular Imaging Spectrometer, OMIS1) remote sensing.

4) Space-borne remote sensing

Compared with other detection techniques mentioned above, the predominant advantage of space-borne remote sensing is the capability to detect coal fires spread out over a large area. During the last decade, research on coal fires in China using space-borne remote sensing data was widely adopted.

2. Controlling Technology of Methane

Methane is one of the greenhouse gases such as carbon dioxide; the global warming potential (GWP) of methane is 21 times higher than that of carbon dioxide. Therefore, if it becomes possible to concentrate methane to an acceptable level for use as fuel for gas engines, boilers, etc., which will greatly contribute to energy conservation and reduction of greenhouse gas emissions. Therefore a methane concentration system is developed which applies the vacuum pressure swing adsorption technology using an **adsorbent**. This adsorbent has a high methane-selectivity against nitrogen and oxygen under a normal atmospheric pressure. The technology is known as the separation technology of gas mixtures. This system can concentrate methane with a concentration of less than 30% emitted to the atmosphere without any effective usage to the concentration above 40%, at which it can be used as fuel for gas engines, boilers, etc.

3. Dust Suppression Techniques

Breathing dust does not become a hazard unless it affects the health or life style of those persons. Dust created in all facets of coal mining can be a health risk and can become an explosion risk.

The proper use of water is the most effective way of controlling dust. Proper use means having an adequate quantity of clean water. Water sprays are useless unless the water supply, quantity and quality are adequate. Cleaning individual water sprays are time-consuming and expensive for coal production. The following points must be followed:

① Ensure quantity and pressure are adequate for the job,

② Install pipes of adequate size,

③ Install and maintain an efficient water filter.

In all cases it is important to have adequate filters or strainers in the water reticulation system. Filters which require personnel to **dismantle** and clean the filter element, then reassemble it, are very rarely cleaned. Filters should be given high priority and consideration should be given to installing reverse flush filters which are easily accessible and can be cleaned without dismantling equipment or even turning off the water supply.

4. Technologies of Preventing Coal Mining Water

The technologies of preventing coal mine water hazards include following four aspects: exploration of hydro-geological conditions; prediction and forecast of water **inrush**; mining under safe water pressure; sealing off groundwater by grouting.

1) Exploration of hydro-geological conditions

Exploration of hydro-geological conditions is essential for preventing coal mine water hazards. Since the mid-term of 1980's, with the need of producing development, there have been two changes on the exploration of hydro-geological conditions: firstly, the research object varied from the large to the small, which was from coal field to mining area and working face. Secondly, exploration engineering transformed from surface to ground.

2) Technology of prediction and forecast of water inrush

It was gradually acknowledged that the water pressure of the Karst aquifer and intensity of protective layer were the main factors that affected water inrush. So the research focused on the coupling of ground pressure, groundwater pressure and the mechanical property and deformation of faults and **fissures**.

Up to the present, the founded theories of coal floor water inrush included "water inrush coefficient theory", "down three zones theory", etc. In addition, system theory, quantity theory, some new methods already were posed and came into operation, such as water inrush information analytic approach, quantity prediction and neural network prediction etc.

3) Technology of mining under safe water pressure

The technology of mining under safe water pressure was referred to as five maps and double coefficients. Five maps included coal floor protection layer broken depth isoline map, coal floor protection layer water pressure isoline map, coal floor aquifer water pressure isoline map, coal floor protection layer thickness isoline map, mining under safe water pressure evaluation map. Double coefficients included water inrush coefficient and safety coefficient of mining under water pressure.

4) Technology of sealing off groundwater by grouting

In the process of sealing off groundwater by **grouting**, in order to increase the amount of cement slurry, controlled orientating drilling technology was applied in some special geological bodies. The special geological bodies were Karst subsidence columns and fissure zones.

In recent years, academician Qian Minggao put forward the concept of green mining and its technical framework. This theory emphasized on alleviating the impact of coal mining mostly. The goal was to maximize the economic and social benefits. The technical framework of green mining included water-preservation in mining areas, coal mining to retard surface subsidence, simultaneous extraction of coal and coalbed methane, reduction of rock waste, underground coal **gasification**, and others.

5) Technology to prevent roof fall

In underground coal mining, room and pillar is one of the oldest methods used for the extraction of flat and tabular coal seams. In this method, a series of rooms are driven in the solid coal using continuous miner and generally shuttle cars and pillars are formed in the development panels. Pillars are left behind to support the roof and to prevent the collapse. To increase the utilization of coal resources during the retreat, the pillars are removed in a later operation (known as retreat mining or pillar recovery). Retreat mining is one of the most hazardous activities because it creates an inherently unstable situation. The process of retreat mining removes the main support for over burden and allows the ground to cave. As a result, the pillar line is an extremely dynamic and highly stressed environment. In other words, the roof at the pillar line is subjected to severe stresses and deformations. Retreat mining accounted for about 10% of all US underground coal production, yet it has historically been associated with more than 25% of all roof and rib fall fatalities between 1986 and 1996.

As a result it can be said that roof controlling is the most challenging safety problem during retreat mining. One of the measures for roof control is sufficient awareness of effective parameters on roof fall. Roof fall depends on a variety of parameters. Each of the parameter may have detrimental effects on miners inform of injury, disability or fatality as well as mining company due to down times, interruptions in the mining operations, equipment breakdowns, etc.

Vocabulary

spatial / ˈspeɪʃ(ə)l / **adj.** in relation to, existing in space 空间的,存在于空间的

spontaneous / spɒnˈteɪnɪəs / **adj.** done, happening, from natural impulse, not caused or suggested by something or somebody outside 自发的,自然的

thermal / ˈθɜːm(ə)l / **adj.** of heat 热的,热量的

adsorbent / ədˈzɔːbənt / **n.** a material, such as activated charcoal, on which adsorption can occur 吸附剂

dismantle / dɪsˈmænt(ə)l / **v.** take away fittings, furnishings, etc. from 拆除,取消,解散

inrush / ˈɪnrʌʃ / **n.** rushing in 涌入,侵入

fissure / ˈfɪʃə / **n.** cleft made by splitting or separation of parts 裂缝,裂沟

grouting / ˈɡraʊtɪŋ / **n.** a thin mortar that can be poured and used to fill cracks in masonry or brickwork (水利)灌浆

gasification / ɡæsɪfɪˈkeɪʃən / **n.** the act of changing into gas 气化

Exercises

Directions: Choose the right answer to each question according to the text.

1. Which of the following is the correct method in fire detection and

UINT 6 Five Major Mining Disasters

monitoring?
   A. adsorbent
   B. water
   C. down three zones theory
   D. space-borne remote sensing
2. How many times of GWP of methane is higher than that of carbon dioxide?
   A. 21
   B. 36
   C. 20
   D. 19
3. What particular element is adopted in the methane concentration system?
   A. water
   B. water filter
   C. adsorbent
   D. pillars
4. Which of the following is NOT the founded theory of coal floor water inrush?
   A. water inrush coefficient theory
   B. down three zones theory
   C. quantity theory
   D. vacuum pressure swing adsorption technology
5. What is considered the most challenging safety problem during retreat mining?
   A. water
   B. roof controlling
   C. methane
   D. dust

Section Three

Extended Reading

   No one ever said that coal mining was easy. In fact, it's outright dangerous. Every day, miners across the country and around the world put themselves on the line. Even with the greatest of safety precautions, accidents happen. Whether it's because of a mechanical or human error, each miner has either experienced or seen his or her share of injuries. When Spike's new original series coal airs Sundays at 8 PM / 7c, the Cobalt Coal crew will share what they've seen and experienced. Until then, here's some insight on what a miner can expect in this highly admirable but difficult life.

Black Lung Disease — The most common term for Coal worker's pneumoconiosis (CWP), black lung disease is caused by long exposure to coal dust. When inhaled, coal dust builds up in the lungs and is unable to be removed by the body, leading to inflammation, fibrosis, and sometimes necrosis, the premature death of cells and living tissue. Since the *Federal Coal Mine Health and Safety Act of* 1969 became law, black lung disease has gone down by 90%, but the downward trend of this disease in coal miners has stopped. Rates are now on the rise, with incidents of black lung disease having doubled in the last 10 years. Reports suggest that close to nine percent of miners with 25 years or more experience tested positive for black lung in 2005-2006, compared with only 4 percent in the late 1990s.

Roof Collapse — Certainly one of the more common causes of injury within coal mines, it's easy to understand why these collapses happen when you realize that you are creating tunnels underground, possibly miles under heavy earth. Roof bolters are responsible for propping up ceilings alongside the thick beams that arch the tunnels themselves, but these devices are never foolproof. Whether it's human error or more simply the settling of the earth, miners always need to be wary of collapse. Even the practice of "retreat mining", which is when miners work backwards, removing beams as they go and allowing the roofs to collapse into an open space, can cause collapses when the next set of supports feels the weight and strain in the process. Despite the best of intentions, certain mine collapses are unpreventable.

Coal Dust Explosions — These explosions are a frequent occurrence and hazard in mines. The explosions occur through the fast combustion of dust particles suspended in the air in an enclosed location. Ignition of the dust particles can happen in any number of ways and a naked flame does not need to present for ignition. Friction, hot surfaces, electrostatic discharge, or simply fire can cause an underground explosion, putting miners' lives in danger, particularly within the confined spaces of mining.

Mining-Induced Seismicity — This refers to typically minor earthquakes and tremors caused by miner activity that can alter the stresses and strains of the Earth's crust. Most seismicity is of an extremely low magnitude, but the sudden shift in rock causes these earthquake-like events that can still collapse mine workings, kill miners, and also damage structures on the surface.

There remains a long list of potential injuries for our coal miners, everything from soft tissue damage, spinal cord injuries, hearing loss, and fractures, not to mention dismemberment. Through safety legislation alongside with improved training practices and technology, mining has become largely safer especially in recent years, leaving deaths and injuries in steep decline, though any level remains unacceptable to the mining community.

Additional words and phrases

exothermic oxidation reaction: 放热氧化反应
coal seam: 煤层
unstable overlying rocks: 不稳定的上覆岩石
coal mine methane (CMM): 煤层气
longwalls faces: 长壁工作面
shear cutting: 剪切
stage loader / crusher: 转载机 / 阶段式破碎机
intake contamination: 摄入污染
coalbed: 煤床
karst water inrush: 岩溶突水
moisture degradation of shale: 页岩水分退化
multiple seam mining: 多煤层开采
underground detecting technique: 地下探测技术
borehole drilling: 深孔凿岩
index gas detection: 索引气体检测
radon gas: 氡气
ground detecting technique: 地面检测技术
self-potential method: 自然电位法
airborne thermal remote sensing: 机载热遥感

unmanned aerial vehicle：无人机
hyper spectral remote sensing：超光谱遥感
water inrush：突水
water inrush coefficient theory：突水系数法
down three zones theory：下三带理论
coal floor protection layer broken depth isoline map：煤层底板保护层破损深度等值线地图
coal floor protection layer water pressure isoline map：煤层地板保护层水压力等值线地图
coal floor aquifer water pressure isoline map：煤层地板含水层水压力等值线地图
coal floor protection layer thickness isoline map：煤层底板防护层厚度等值线图
mining under safe water pressure evaluation map：矿业安全水压力评价地图
safety coefficient of mining under water pressure：水压力下采煤安全系数
green mining：绿色开采
retreat mining：撤退式采煤

Unit 7 Ecological Crisis

Foreword

For quite a long time, coal has been used as a source of energy by mankind. Pertaining to power generation, coal is one of the most cost-effective fossil fuels. Not only is the electricity generated from coal enormously, but it is also more reliable than other sources of energy. Although it brings benefits to our prosperity, the burning of coal has caused many ecological problems such as global warming, climate change, and bio-diversity extinction. When coal is being burned, harmful gases such as ash, sulphuric dioxide and carbon dioxide are released into the air. Actually, coal gives off twice as much carbon dioxide than any other fossil fuel. Therefore, we must find ways to offset the harm coal does to the environment to strike a balance between coal exploration and social sustainable development.

Section One

Preview

Text A tells us that our world is now facing the prospect of an ecological collapse, which is not sensational but based on the fact that the way we develop economy is unsustainable. What makes the current ecological situation so serious is the climate change, which arises from human-generated increases in greenhouse gas emissions. To what extent will this climate change affect the future of humanity? How does the world respond to the present ecological crisis? What strategies should we take to address the problem? As you read, you will hopefully find answers to the above questions.

Text A

Are We Coming Close to an Ecological Collapse?

It is now universally recognized within science that humanity is confronting the prospect — if we do not soon change course — of a **planetary** ecological collapse. Not only is the global ecological crisis becoming more and more severe, with the time in which to address it fast running out, but the dominant environmental strategies are also forms of denial, **demonstrably** doomed to fail, judging by their own limited objectives. This tragic failure can be attributed to the refusal of the powers that are to address the roots of the ecological problem in capitalist production and the resulting necessity of ecological and social revolution.

The term "crisis," attached to the global ecological problem, although unavoidable, is somewhat misleading, given its dominant economic associations. But insofar as it is related to the business cycle and not to long-term factors, expectations are that it is temporary and will end, to be followed by a period of economic recovery and growth — until the **advent** of the next crisis. Capitalism is, in this sense, a crisis-ridden and cyclical economic system.

When we speak today of the world ecological crisis, however, we are referring to something that could turn out to be final, i.e., there is a high probability, if we do not quickly change course, of a terminal crisis — a death of the whole **anthropocene**, the period of human dominance of the planet. Human actions are generating environmental changes that threaten the extermination of most species on the planet, along with civilization, and conceivably our own species as well.

What makes the current ecological situation so serious is that climate change, arising from human-generated increases in greenhouse gas emissions, is not occurring gradually and in a linear process, but is undergoing a dangerous acceleration, pointing to sudden shifts in the state of the earth system.

Due to this acceleration of climate change, the time line in which to act before **calamities** hit, and before climate change increasingly escapes our control, is extremely short. It was reported that, based on current trends, close to 70 percent of the land surface of the earth could be drought-affected by 2025, compared to nearly 40 percent today. Warnings are given that Himalayan **glaciers** could disappear altogether by 2035. Rivers fed by these glaciers currently supply water to over half the world's population. Their melting will give rise to enormous floods, followed by acute water shortages.

Many of the planetary dangers associated with current global warming trends are by now well-known: rising sea levels engulfing islands and lowlying coastal regions throughout the globe; loss of tropical forests; destruction of coral reefs; a "sixth extinction" rivaling the great **die-downs** in the history of the planet; massive crop losses; extreme weather events; spreading hunger and disease. But these dangers are heightened by the fact that climate change is not the entirety of the world ecological crisis.

Since the whole earth is affected by the vast scale of human impact on the environment in complex and unpredictable ways, even more serious **catastrophes** could **conceivably** be set in motion. One growing area of concern is ocean **acidification** due to rising carbon dioxide emissions. Within a decade, the waters near the North Pole could become so corrosive as to dissolve the living shells of shellfish, affecting the entire ocean food chain. At the same time, ocean acidification appears to be reducing the carbon uptake of the oceans and speeding up global warming.

There are endless predictive uncertainties in all of this. Nevertheless, evidence is mounting that the continuation of current trends is **unsustainable**, even in the short-term. The only rational answer, then, is a radical change of course. Moreover, given certain **imminent tipping** points, there is no time to be lost. Catastrophic changes in the earth system could be set **irreversibly** in motion within a few decades, at most.

Some international organizations reported that an atmospheric carbon dioxide level of 450 parts per million (ppm) should not be exceeded, and implied that this was the **fail-safe** point for carbon stabilization. But these findings are already out of date. The updated news shows that carbon emissions have to be reduced faster and more drastically than originally thought, to bring the overall carbon concentration in the atmosphere down. The reality is that, "if we burn all the fossil fuels, or even half of remaining reserves, we will send the planet toward the ice-free state with sea level about 250 feet higher than today. It would take time for complete ice sheet **disintegration** to occur, but a chaotic situation would be created with changes occurring out of control of future generations."

The central issue that we have to confront, therefore, is devising social strategies to address the world ecological crisis. Not only do the solutions have to be large enough to deal with the problem, but also all of this must take place on a world scale in a generation or so. The speed and scale of change necessary means that what is required is an ecological revolution that would also need to be a social revolution. However, rather than addressing the real roots of the crisis and drawing the appropriate conclusions, the dominant response is to avoid all questions about the nature of our society, and to turn to technological fixes or market mechanisms of one sort or another. In this respect, there is a certain continuity of thought between those who deny the climate change problem altogether, and those who, while acknowledging the severity of the problem at one level, nevertheless deny that it requires a revolution in our social system.

We are increasingly led to believe that the answers to climate change are primarily to be found in new energy technology, specifically increased energy and carbon efficiencies in both production and consumption. Technology in this sense, however, is often viewed abstractly as a **deus ex machina**, separated from both the laws of physics (i.e., **entropy** or the second law of **thermodynamics**) and from the way technology is **embedded** in historically specific conditions. With respect to the latter, it is worth noting that, under the present economic system, increases in energy efficiency normally lead to increases in the scale of economic output, effectively **negating** any gains from the standpoint of resource use or carbon efficiency — a problem known as the "**Jevons Paradox.**"

Technological **fetishism** with regard to environmental issues is usually coupled with a form of market fetishism. Green-market fetishism is most evident in what is called "**cap and trade**" — a catch phrase for the creation, via governments, of artificial markets in carbon trading and so-called "offsets." The important thing to know about cap and trade is that it is a proven failure.

The **masquerade** associated with the dominant response to global warming is illustrated in the climate bill passed by the U. S. House of Representatives in late June 2009. The bill, if **enacted**, would supposedly reduce greenhouse gas emissions 17 percent relative to 2005 levels by 2020, which translates into 4-5 percent less U. S. global warming pollution than in 1990. This then would still not reach the target level of a 6-8 percent cut (relative to 1990) for wealthy countries that the **Kyoto accord** set for 2012, and that was supposed to have been only a minor, first step in dealing with global warming — at a time when the problem was seen as much less severe. The goal presented in the House bill, even if reached, would therefore prove vastly inadequate.

But the small print in the bill makes achieving even this **meager** target unrealistic. The coal industry is given until 2025 to comply with the bill's pollution reduction **mandates**, with possible extensions afterward, the bill, however, builds in approval of new coal-fired power plants!

Recognizing that world powers are playing the role of Nero as Rome burns, some experts argue that massive climate change and the destruction of human civilization as we know it may now be irreversible.

Rational scientists recognize that interventions in the earth system on the scale **envisioned** by **geoengineering** schemes (for example, blocking sunlight) have their own massive and unforeseen consequences. Nor could such schemes solve the crisis. The dumping of massive quantities of sulfur dioxide into the **stratosphere** would, even if effective, have to be done again and again, on an increasing scale, if the underlying problem of cutting greenhouse gas emissions were not dealt with. Moreover, it could not possibly solve other problems associated with massive carbon dioxide emissions, such as the acidification of the oceans.

The dominant approach to the world ecological crisis, focusing on technological fixes and market mechanisms, is thus a kind of denial; one that serves the **vested** interests of those who have the most to lose from a change in economic arrangements. Al Gore exemplifies the dominant form of denial in his new book, *Our Choice: A Plan to Solve the Climate Crisis*. For Gore, the answer is the creation of a "sustainable capitalism." He is not, however, altogether blind to the faults of the present system. He describes climate change as the "greatest market failure in history" and **decries** the "short-term" perspective of present-day capitalism, its "market triumphalism," and the "**fundamental flaws**" in its relation to the environment. Yet, in **defiance** of all this, he assures his readers that the "strengths of capitalism" can be **harnessed** to a new system of "sustainable development."

Vocabulary

planetary / ˈplænɪt(ə)rɪ / *adj.* relating to or belonging to planets 行星的
demonstrably / ˈdemənstrəblɪ / adv. 显而易见地
advent / ˈædvənt / *n.* arrival that has been awaited (especially of something momentous) 出现
anthropocene / ænˈθrɒpəˌsiːn / *n.* a proposed term for the present geological epoch (from the time of the Industrial Revolution onwards), during which humanity has begun to have a significant impact on the environment 人类世；人类自工业革命以来的活动对环境的影响可成立一个新地质时代的理论
calamity / kəˈlæmɪtɪ / *n.* an event that causes a great deal of damage, destruction, or personal distress 灾难
glacier / ˈglæsɪə; ˈgleɪsɪə / *n.* an extremely large mass of ice which moves very slowly, often down a mountain valley 冰川
die-down *n.* 逐渐消失
catastrophe / kəˈtæstrəfɪ / *n.* an unexpected event that causes great suffering or damage 灾难
conceivably / kənˈsiːvəblɪ / adv. 可想像地，可相信地
acidification / əˌsɪdɪfɪˈkeɪʃən / *n.* the process of becoming acid or being converted into an acid 酸化
unsustainable / ˌʌnsəˈsteɪnəb(ə)l / *adj.* that cannot be continued at the same level, rate, etc 不能持续的；无法维持的
imminent / ˈɪmɪnənt / *adj.* likely to happen very soon （尤指不好的事情）即将发生的
tipping *adj.* 倾斜的
irreversibly / ɪrɪˈvɜːsɪb(ə)lɪ / adv. 不可逆地，无法挽回地
fail-safe / ˈfeɪlseɪf / *adj.* Something that is fail-safe is designed or made in such a way that nothing dangerous can happen if a part of it goes wrong 自动防故障装置的
disintegration / dɪsˌɪntɪˈgreɪʃ(ə)n / *n.* 解体，瓦解，分崩离析

Unit 7　Ecological Crisis

entropy / ˈentrəpɪ / *n.* a measurement of the energy that is present in a system or process but is not available to do work 熵（物质系统的不能用于做功的能量的度量）

thermodynamics / ˌθɜːmədaɪˈnæmɪks / *n.* the branch of physics that is concerned with the relationship between heat and other forms of energy 热动力学

embed / ɪmˈbed / *v.* to fix something firmly into a substance or solid object 嵌入

negate / nɪˈɡeɪt / *v.* to stop something from having any effect 取消，使无效；to state that something does not exist 否定；否认

paradox / ˈpærədɒks / *n.* a situation when it involves two or more facts or qualities that seem to contradict each other 自相矛盾

fetishism / ˈfiːtɪʃɪzəm / *n.* a strong liking or need for a particular object or activity which gives people sexual pleasure and excitement 恋物癖

masquerade / ˌmɑːskəˈreɪd; ˌmæskəˈreɪd / *n.* an attempt to deceive people about the true nature or identity of something 伪装；*v.* to pretend to be a person or thing, particularly in order to deceive other people 冒充

enact / ɪˈnækt / *v.* to put something into action, esp. to change something into a law 制定，通过；to perform (a part in a play) 扮演，演出

accord / əˈkɔːd / *n.* a formal agreement, for example, to end a war between countries or groups of people 协议

meager / ˈmiːɡə / *adj.* very small or not enough 不足的，微薄的，贫乏的

mandate / ˈmændeɪt / *n.* the authority to do something, given to a government or other organization by the people who vote for it in an election 授权

envision / enˈvɪʒ(ə)n / *v.* to imagine what a situation will be like in the future, especially a situation you intend to work towards 展望，想象

geoengineering *n.* 地球工程学

stratosphere / ˈstrætəˌsfɪə / *n.* the layer of the Earth's atmosphere which lies between 7 and 31 miles above the earth 平流层；同温层

vest / vest / *v.* to bestow a power on somebody or something 授予

decry / dɪˈkraɪ / *v.* to strongly criticize somebody or something, especially publicly （公开）谴责；（强烈）批评

defiance / dɪˈfaɪəns / *n.* open refusal to obey somebody or something 反抗；违抗；拒绝服从

harness / ˈhɑːnɪs / *v.* to gain control of something and use it for some purpose 控制；利用；治理

Phrases and Expressions

pertain to：关于，属于
attribute ... to ...：认为……是……的结果
insofar as：到……程度
give rise to：引起，导致
tipping point：临界点；拐点；转折点
with respect (regard) to：关于，至于
from the standpoint of：从……角度（立场）来看
be coupled with：伴随，与……相结合
catch phrase：口头禅，标语
be relative to：相对于
translate into：转化成，转变为
comply with：遵守，遵从
vested interests：既得利益
be blind to：对……视而不见
in (one's) relation to：关于，涉及
in defiance of：不顾，无视

Proper Words and Terms

deus ex machina / ˈdeɪəs eks ˈmɑːkiːnə / a calque from Greek, meaning "god from the machine". The term has evolved to mean a plot device whereby a seemingly unsolvable problem is suddenly and abruptly resolved by the contrived and unexpected intervention of some new event, character, ability or object. Depending on how it is done, it can be intended to move the story forward when the writer has "painted himself into a corner" and sees no other way out, to surprise the audience, to bring the tale to a happy ending, or as a comedic device. 解围的人或事件；大突破；杀出重围

Jevons / ˈdʒevənz / **Paradox**: is the proposition that as technology progresses, the increase in efficiency with which a resource is used tends to increase (rather than decrease) the rate of consumption of that resource 杰文斯悖论：

经济现象,某自然资源的消耗,会随利用该资源之技术的改进而加快,因为技术改进会使下游产品价格降低,进一步刺激人们对该产品的需求。

cap and trade: one method for regulating and ultimately reducing the amount of pollution emitted into the atmosphere. It is viewed as a more democratic solution to regulating pollution than a carbon tax as it creates a commodity out of the right to emit carbon and allows the commodity to be traded on the free market. 总量管制和交易制度,碳排放权交易制度(要求那些温室气体的排放者接受或购买排放许可,可获得的许可的数量将会逐渐减少,以此来限制温室气体排放,由美国 Waxman-Markey 提案提出。)

Kyoto / kɪˈəutəu / a city located in the central part of the island of Honshu, Japan. Formerly the imperial capital of Japan for more than one thousand years, it is now the capital city of Kyoto Prefecture located in the Kansai region(关西地区), as well as a major part of the Kyoto-Osaka-Kobe(京都-大阪-神户)metropolitan area. One historical nickname for the city is the City of Ten Thousand Shrines.

Exercises

Ⅰ. Comprehension of the Text

Directions: Please answer the following questions according to the text.

1. What is the world now faced with?
2. What do environmental changes threaten?
3. How much of the world land surface could suffer from drought by 2015?
4. What is the direct cause of ocean acidification?
5. What is the answer to the current unsustainable development?
6. Does the writer prefer the technological approach to deal with climate change? Why or why not?
7. Why does the writer compare the dominant response to global warming to a masquerade?
8. What does the writer think of the dominant approach to the world ecological crisis if people only focus on technological fixes and market mechanisms?

Ⅱ. Group Discussion

Directions: Discuss the following questions with your partners, using as

much text information as possible in your discussion.
1. Do you think human beings can harness the climate change? Why or why not?
2. Talk about technological fetishism and market fetishism by citing some examples.
3. How do you think of the current exploration and use of coal, one of fossil fuels? What suggestions can you propose as to the efficient and green use of coal?

Ⅲ. Word Bank

Directions: *Fill the blanks with the words on the right side of the text. For each word, you can use only once.*

Humanity is now confronted with an ecological 1. _____, but the dominant environmental strategies are just forms of 2. _____. The most worrisome environmental situation is the climate change, which arises from 3. _____ increases in 4. _____ gas emissions. To cope with the problem resulting from the human actions, the writer thinks the central issue is to come up with social strategies to 5. _____ the world ecological crisis. However, the dominant response to crisis is to 6. _____ all the questions about the nature of our society, and turn to technological 7. _____ or market 8. _____ of one sort or another, which the writer thinks are forms of 9. _____. Rational scientists recognize that interventions in the earth system on the scale 10. _____ by geoengineering schemes have their own massive and unforeseen consequences.

a) greenhouse
b) fixes
c) stratosphere
d) collapse
e) fetishism
f) denial
g) human-generated
h) envisioned
i) avoid
j) address
k) masquerade
l) mechanisms
m) global
n) imminent
o) conceivable

Ⅳ. Translation Practice

Directions: *Translate the following sentences from Chinese to English.*
1. 煤炭价格谈判失败的主要原因是发电企业储煤量充足。(be attributed to)
2. 就全球变暖而言,那是由于化石燃料燃烧时所释放的二氧化碳造成的。

(insofar as)
3. 越来越多的证据表明,目前的经济发展模式是不可持续的。(Evidence is mounting that ...)
4. 我们必须在世界范围内寻找解决降低空气中二氧化碳含量的策略。(on a world scale)
5.《环境保护法(修订)》在此次人大会议上全票通过。(enact)
6. 现实情况是,煤炭在中国能源消费结构中占主要地位。(The reality is ...)

Ⅴ. Writing Practice

Directions: Write a 3-paragraph passage of about 120 words with the title "My View on the Future of Coal Utilization" based on the outline below.
1. The present situation and role of coal use.
2. The use of coal has given rise to many problems.
3. My view on the future of coal use.

Section Two

Preview

Text B takes for example two countries, Spain and the Netherlands, to give us a clear and true picture of renewables in energy consumption structure. According to the author, the so-called greening of our electricity production is still 100 percent wishful thinking. Thus, the energy crisis is worsening instead of improving. It's now high time to look at energy use in a new angle.

Text B

How (Not) to Resolve the Energy Crisis

Increasing the share of renewable energy will not make us any less dependent on fossil fuels as long as total energy consumption keeps rising. Renewable energy sources do not replace coal, oil or gas plants, for they only meet (part of) the growing demand.

Regardless of the growing share of renewable energy sources, we burn up more and more fossil fuels every year. This is the case in the US, in Europe and on a global scale, but to make my point I will start by analyzing the situation in Spain and in the Netherlands, because both countries are regarded to be an example for their commitment to renewable energy.

Moreover, the Netherlands have a **negligible** share of nuclear energy and hydropower, while in Spain these energy sources have remained unchanged over the last decade, which makes the calculations more clear.

Share of Renewables

Spain once made headlines around the world with the news that it generated over 53 percent of its electricity by wind power alone, be it during an extremely windy night and only for some hours. There is no denying that the development of wind power in Spain is impressive. Electricity generated by wind power grew with 8,000 percent between 1996 and 2007, from 338 gigawatt-hours (GWh) in 1996 to 27,509 GWh in 2007. With it, the share of wind power in electricity production grew from 0.2 to 9 percent. In the Netherlands, the amount of "green" electricity increased by 400 percent between 1998 and 2008, from 2,300 GWh to 9,500 GWh. With it, the share of renewable energy (mostly **biomass** and wind) in electricity production grew from 2.5 percent to 9 percent.

This sounds great, especially when you compare it to the situation in the United States, where the share of renewable energy in electricity production (excluding hydropower) rose from 1.4 percent to 2.3 percent during the same period (1996 — 2007). Or, on a global scale, where the share of renewables rose from 1.12 percent in 1990 to 2.3 percent in 2006. Yet, just like the Americans and the rest of the world, the Spanish and the Dutch are now *more* dependent on fossil fuels than a decade ago, not less.

Total Electricity Production

The reason is, of course, that the total electricity production in both countries kept rising. In Spain, it went up from 174,246 GWh in 1996 to 303,293 GWh in 2007 (a rise of almost 80 percent in 11 years). The share of fossil fuels in electricity generation grew from 38 percent in 1996 to 59 percent in 2007, while the absolute amount of fossil fuels used for electricity generation grew from 67,651 GWh to 179,737 GWh. So, from 1996 to 2007 the amount of wind powered electricity in Spain grew with 27,171 GWh, and the amount of fossil fuel powered electricity grew with 112,086 GWh. Now please explain to me, what is so "green" and exciting about this trend?

Avoided Emissions

Of course, things could have been even worse: that is why policymakers and **statisticians** prefer to talk about "avoided use of fossil energy" and "avoided CO_2-emissions". The reasoning goes as follows: if we would not have built those wind **turbines** and solar panels, then we would have burnt up *even more* fossil fuels. But, who are we fooling here?

The Spanish would have "avoided" the same amount of emissions and fossil energy if they would have built *not one wind turbine* between 1996 and 2007, but had chosen to limit the rise of energy consumption to 84,915 GWh, instead of the recorded 112,086 GWh. If they would have done that, they would have been just as dependent on fossil fuels as they are today, and they would have emitted the same amount of greenhouse gases as they do today — all this without those 27,171 GWh of wind powered electricity. They would not have made headlines with it, though. Nobody would have noticed.

The same goes for the Dutch: they would have "avoided" the same amount of emissions and fossil energy if they would have limited the rise of energy consumption to 7,000 GWh, instead of the recorded 14,000 GWh.

Embodied Energy

In fact, this low-tech **scenario** would have been a more ecological and energy-efficient choice, because both countries would have saved the energy required to produce those renewable energy plants and sources — solar panels, wind turbines and wood **pellets**. Green electricity is not generated by a "clean" energy source, but by a "cleaner" energy source. Solar panels, wind turbines and wood pellets do not use gas or coal during their operation, but they do require energy for their production (and since they are mostly produced far away from the place where they are used these figures do not show up in national statistics of energy consumption).

Mind you: the embodied energy of wind turbines (and solar panels) is not a problem if they replace non-renewable energy plants, because in that case we do save energy and thus make progress. But, this is not the case, so the embodied energy of this added electricity generation capacity is definitely extra energy use.

We Do too Much

This does not mean that coal plants are preferable to wind turbines and solar panels. In fact, if the Dutch had built the (7,200 GWh) renewable energy plants and not built the (7,000 GWh) non-renewable energy plants, the result would have been real progress. Likewise, if they would have frozen energy consumption at the 1998 level and built nothing — renewable nor non-renewable energy plants — again there would have been substantial progress. They would be less dependent on fossil fuels and they would produce less CO_2 and air pollution.

The problem is that they did not do any of this. Or, better said, they did everything at the same time; constructing more renewable energy plants, constructing more non-renewable energy plants, and consuming more energy.

Piling up Energy Sources

Again, this trend is not limited to Spain and to the Netherlands, and what is happening is not a new phenomenon either. What we are doing for more than 100 years now, is piling up energy sources. Today (in the Netherlands, Spain, the US and worldwide) the absolute amount of coal consumed for electricity production is much larger than one century ago, when there was no talk of gas, oil and nuclear. The dirty coal of the beginning of the industrial revolution was not replaced by cleaner gas plants. The gas plants joined the coal plants.

Next, nuclear plants did not replace the existing coal and gas plants, instead they joined them. Today, with renewable energy, the same thing is happening. They address an energy demand that did not exist before. We use renewable energy sources to power an ever growing **plethora** of energy-sucking gadgets — and this will not get us anywhere.

Up until now, newer and cleaner energy sources have always been used to enlarge energy production, not to make it "greener".

The so-called greening of our electricity production, which generates so much talk, is still 100 percent **wishful** thinking. We are not one step further than 5, 10, 20 or even 100 years ago. On the contrary, things get worse every day.

Relative Versus Absolute Figures

Much more important than what we do, is what we don't do. The key to progress is scaling down non-renewable energy production, or at least keeping it at the same level. Instead of aiming for the development of more renewable energy, policymakers should do anything in their power to make sure that not one more kilowatt of non-renewable energy is added.

Problem is that all policy objectives are expressed in relative terms (as a percentage of total electricity production) and never as absolute figures. The European Union aims to generate 20 percent of total electricity production by wind energy and 15 percent by solar energy in 2020. The US aims to generate 25 percent of its electricity from renewable sources by 2025. None of their reports describe any goal in absolute figures. This is a fruitless approach as long as total electricity consumption is on the rise.

How to Solve the Energy Crisis

Don't get me wrong: all efforts to build and develop renewable energy and energy efficient technology are useful and very necessary. My point is that, by themselves, they will not yield any results. To make them work, we need to put an absolute limit to energy *use*.

Imagine that the European Union or the US would decide that in 2020 we can only use as much energy (or electricity) as we do today. Interestingly, all other efforts suddenly make sense. If the share of renewable energy would rise, then the share of non-renewable energy would automatically fall. Energy efficient technology would be automatically transformed in energy savings, and not in extra applications or performance, as it happens now (the energy efficiency paradox).

In this scenario, with every small step forward in renewable energy production and energy efficient technology, we would become less and less dependent on fossil fuels, and we would emit less and less greenhouse gases. Moreover, it is hard to call this measure drastic or radical: if we can manage today with 18,008,000 Gwh, why not in 2030? What more energy-sucking gadgets do we need? By the way, we can keep developing new products and services as we please, making sure that these are energy efficient. Considering the amount of energy that is now wasted by most products and services, there is lots of room for innovation, improvement, and thus economic growth.

Right now we try to match our energy production to an ever increasing demand. But, we could also try to match our demand to a fixed supply. Considering the circumstances, this would be a much more realistic and intelligent strategy. An even better strategy is the "oil depletion protocol", an idea of author Richard Heinberg. He proposes an international agreement to *lower* oil production and consumption each year with 2.6 percent. We can wait until the geological, economical or geopolitical reality lowers the availability of fossil fuels, but if we anticipate that reality now then we definitely have more of a chance to make a successful transition to a durable and less energy-intensive society.

Not China's Fault

Last, but not least, the IEA notes that the rise of energy use is largely on account of non-western countries, with China ahead. But, this does not clear us at all. As the IEA calculated in a former report, almost 30 percent of energy use in China comes from the production of export goods — from bicycles over jeans to solar panels.

Western countries succeed in limiting the rise of their energy consumption because they have **outsourced** ever more energy use. Moreover, the IEA states in its last report, non-OECD countries are, in spite of their high share in current energy use, only responsible for 42 percent of the CO_2-emissions since 1890 — with a much bigger population. This means that — in a fair world — we would have to reduce our energy use much more than them.

Vocabulary

negligible / ˈneglɪdʒəb(ə)l / *adj.* of very little importance or size and not worth considering; insignificant 微不足道的;不重要的;不值一提的

biomass / ˈbaɪəʊmæs / *n.* the total quantity or weight of plants and animals in a particular area or volume (单位面积或体积内的)生物量

statistician / ˌstætɪˈstɪʃ(ə)n / *n.* a person who studies or works with statistics 统计学家;统计员

turbine [ˈtɜːbaɪn] *n.* a machine or an engine that receives its power from a wheel that is turned by the pressure of water, air or gas 涡轮机;汽轮机

scenario / səˈnærɪəʊ / *n.* a description of how things might happen in the

Unit 7 Ecological Crisis

future; a written outline of what happens in a film / movie or play 设想,方案,预测;(电影或戏剧的)剧情梗概

pellet / ˈpelət / **n.** a small hard ball of any substance, often of soft material that has become hard 小球;团粒;丸

plethora / ˈpleθərə / **n.** an amount that is greater than is needed or can be used 过多;过量;过剩

wishful / ˈwɪʃfʊl / **adj.** having wishes or characterized by wishing 渴望的,愿望的;寄予希望的

outsource / ˈaʊtsɔːs / **v.** to arrange for somebody outside a company to do work or provide goods for that company 交外办理;外购;外包

Exercises

Directions: Choose the right answer to each question according to the text.

1. According to the text, which of the following countries consumes an increasing amount of fossil fuel?
   A. U.S.A.
   B. Netherlands
   C. Europe
   D. All of the above

2. Which of the following is true according to the text?
   A. Wind power developed in Spain is negligible.
   B. Green electricity production in the Netherlands has decreased in the past decade.
   C. The share of renewable energy in electricity production in the USA didn't rise as much as in Spain.
   D. Spain and the Netherlands are at present more independent of fossil fuels.

3. What does the writer say about the share of fossil fuels in electricity generation in Spain?
   A. It increased.
   B. It decreased.
   C. It kept unchanged.
   D. Not mentioned.

4. According to the writer, the production of renewable energy plants and

sources _____ .
A. is totally green
B. needs non-fossil fuels
C. needs fossil fuels
D. only needs coal

5. From the text, it can be seen that the writer _____ .
A. wishes the Dutch could build renewable energy plants
B. does not wish the Dutch to build renewable energy plants
C. wishes the Dutch could build non-renewable energy plants
D. does not wish the Dutch to build non-renewable energy plants

Section Three

Extended Reading

Why Introduce Yet Another Expensive, Energy-intensive and Risky Technology?

Capturing CO_2 from the smokestacks of power stations with the intention of storing it in underground reservoirs, oceans, rocks, consumer products, chemicals or fuels has gained a lot of credibility recently. Many experts believe that we will burn the world's remaining fossil fuels anyway, and we should therefore try to lower the impact if we are to prevent a catastrophic climate change. Yet capturing, transporting and storing carbon dioxide raises energy consumption considerably and brings with it serious health and environmental problems. The benefits, on the other hand, are shadowed in doubt.

Earlier this month leading science and energy institutes advocated strongly for the development of carbon capture and storage technology. The science academies of the world's 13 major economic powers called the implementation of carbon sequestration a "top priority". Around the same time, the International Energy Association (IEA) argued for an energy technology revolution, of which carbon capture and storage forms a vital component. Meanwhile, many spin-offs and start-ups are presenting all kinds of "innovative" ideas that seem to differ substantially from the traditional approach of storing CO_2 in underground reservoirs.

The idea of carbon capture and storage (CCS) — first introduced in the 1970s — is attractive at first sight. To begin with, it is a natural occurrence. There are many natural reservoirs of CO_2, which have kept this gas contained for many millions of years. Secondly, the potential for storage is significant. The available storage space in underground reservoirs (depleted oil and gas reserves, coal formations and especially saline formations) is probably large enough to store all the carbon dioxide still contained in earth's remaining fossil fuel reserves. It will take more research to find out which reservoirs are suited to this and which ones are not, but finding the actual storage space does not seem to be a fundamental obstacle. To add to this, the technology for capture, transport and storage of CO_2 is available.

CO_2 can be transported in gaseous, liquid and solid form. The latter is called dry ice. It can be kept at a temperature of minus 70 degrees Celsius and transform directly to gas when it melts (hence the name). One cubic meter of dry ice equals to 1.5 tons of carbon dioxide, and in gaseous form the same amount of CO_2 takes up more than 500 m^3. For transportation in pipelines, CO_2 gas is compressed. For transportation by ships, CO_2 can be transported in any form.

Capturing CO_2 from smokestacks has been a common practice for many years. Injection and storage of carbon dioxide is already happening in the North Sea, in Algeria and in Texas. In these cases, CO_2 is injected into oil and gas reservoirs in order to extract more fossil fuels than would otherwise be possible, a process called Enhanced Oil Recovery (EOR). For some of these applications, carbon dioxide is transported by pipeline or by ship.

A complete CCS infrastructure has not been demonstrated yet (all CO_2 used for enhanced oil recovery is commercially produced or originates from other sources than power plants, and present capture techniques do not capture CO_2 for storage but emit the gas in the atmosphere). Yet, since all the individual parts exist, this does not seem to be an obstacle either.

The problem at hand is that the process of capturing, transporting and storing carbon dioxide requires a vast amount of energy. If this energy were to be derived directly from fossil fuels the benefits of the CO_2-savings by capture and storage would be offset by the very same energy intensive process. If the energy were to come from renewable sources the technology would be rendered unnecessary as it would be much more efficient to generate electricity directly from the renewable source.

Capturing CO_2 from smokestacks is the most energy-intensive part of the process. According to the International Panel of Climate Change (IPCC), which devoted a comprehensive study on the technology 3 years ago, capturing technology (including compression for further transport and storage) raises the energy consumption of a coal plant by an average of 32 percent.

A coal plant equipped with CO_2-capture technology would thus need 32 percent more coal and other resources like water, chemicals and reagents to produce the same amount of electricity than the same power plant without this technology. Carbon capture technology forms a symbiosis with coal as a fuel ("clean coal"), since burning coal emits twice as much greenhouse gasses than burning gas. Capturing CO_2 from a gas power plant requires less energy but is of not much use.

This 32 percent does not include the energy needed to mine, process and transport of the many thousands of tons of extra coal, and it does not include the energy needed for the construction of the capture, transportation, storing and monitoring infrastructure either.

It is insufficient to simply place the smokestacks of a coal plant upside down as suitable underground reservoirs do not necessarily lie beneath the world's power stations. A carbon capture and storage infrastructure requires a transport infrastructure consisting of pipelines (and tankers) that rivals the existing oil and gas network.

Manufacturing and installing these thousands of kilometers of stainless steel pipes will require a substantial amount of energy. Also, the transport by ship or pipeline itself requires energy, and so does the injection of the CO_2 in underground reservoirs and the monitoring of the whole transport network (today's pipelines are patrolled by plane every two weeks).

Everything taken together, CCS will probably raise energy consumption by as much as 50 percent (this is an estimate as to my knowledge nobody seems to have investigated this yet). Even if all of the CO_2 is eventually stored, a 50 percent increase in energy consumption is the last thing that the world needs.

In addition, it is impossible to store all of the carbon dioxide. Capturing technology can only capture 80 to 90 percent of CO_2 from smokestacks. Taking into account an additional energy use of 50 percent this comes down to a reduction in CO_2 emissions of only 70 to 85 percent compared to a coal plant without carbon capture technology (and only if all emissions coming from the additional energy use are captured as well).

There are losses during transport, too. According to the IPCC these are 1 to 2 percent per 1,000 kilometers of pipeline transport and 3 to 4 percent per 1,000 kilometers of ship transport (the ship's fuel use included). Carbon dioxide is also not the only harmful effect of power generation. Burning coal brings with it serious air pollution and produces waste, both of which will also rise by at least 30 percent. The same goes for the ecological damage of coal mining, which is devastating. Storing the CO_2 can never prevent this.

Because of all these disadvantages, researchers and entrepreneurs try to invent all kinds of other ways to keep carbon dioxide out of the atmosphere, like storing CO_2 in consumer products, and converting it to fuel or fixing it in rocks. Yet until now, all these proposals face — at best — a similar energy penalty.

That is mostly because the first step of the process is always the same: capturing CO_2 from the smokestacks of a coal plant will raise fuel consumption by about 30 percent. The only alternative to this capture technology — sucking carbon dioxide out of ambient air by means of "artificial trees" — consumes even more energy because the concentration of CO_2 in ambient air is much lower than the concentration of CO_2 in smokestacks.

Besides, the storage potential of most of these alternative proposals is very limited compared to that of the traditional CCS-concept. A strategy that is getting a lot of attention these days is to store CO_2 into consumer products based on polymers like plastic bottles or DVDs, or in chemicals like those used in fertilizers, refrigeration or food packaging.

The idea is that it is better to do "something useful" with captured CO_2 instead of just storing it underground. However, even though the amount of chemicals and plastics we produce is enormous, as a carbon sink they are all but meaningless.

According to the IPCC, producing all polycarbonates and polyurethane by means of carbon dioxide would store 3.3 million tons of CO_2 — comparable to the annual emissions of just 3 coal power plants. China is building one coal plant per week (in large part to produce cheap goods for us to buy) and there are more than 100 coal plants on the drawing board in the US.

What makes this approach even more useless is that these consumer products and chemicals have a relatively short lifespan, from a few months for fertilizers to a few decades for plastic products. When the fertilizers are used, or when the DVDs end up in the incinerator, the CO_2 goes back into the atmosphere.

Burning Algae

Using CO_2 as a feeding stock for algae and then turning it into biofuel — another idea that is hyped these days — faces the same problem. It only delays CO_2-emissions for a very short time. The carbon dioxide is converted into fuel which is subsequently burned in a car engine.

It is impossible to capture CO_2 from car engines since the gas is too heavy (your car would gain serious weight while driving, and it would have to pull a trailer to store the large volume of carbon dioxide).

One could argue that at least the CO_2 is recycled and that we are using the by-product of electricity generation to make fuel — which means that we don't have to dig up more fossil fuels to make gasoline. However, this argument does not take into account the fact that much energy (and water) is lost in the conversion process.

Firstly, there is again the energy penalty of CO_2 capture from the smokestacks, on average 30 percent. Next, you have to build a huge infrastructure to produce algae (since their energy efficiency is 100 times smaller than that of solar panels) and furthermore there is the energy that gets lost during the process of turning algae to fuel. If there is net energy gain in the end, it will be small.

Turning CO_2 into algae could be an interesting strategy to reduce CO_2-emissions if we store the algae underground instead of burning them in our car engines (thus avoiding the energy-intensive process of converting them into fuel). However, that's not on anyone's mind.

The only alternative approach that really could store a significant amount of CO_2 is mineral carbonation: fixing carbon dioxide in rocks (limestone).

This method is also based on a natural process. Under the influence of weathering, the surface of rocks combines with carbon dioxide. Earth has no shortage of rocks, so the potential of this method is large enough to store all possible future CO_2-emissions.

But if we want the strategy to be useful for us, this natural chemical reaction has to be accelerated considerably and that is again a very energy-intensive process (rocks have to be crushed to powder in order to provide more rock surface, and then treated with chemicals and heated). According to the IPCC this method would raise energy consumption by 60 to 180 percent.

Capturing carbon dioxide in rocks would also require a mining and transport infrastructure that is comparable (and which would supplement) to today's coal industry. To fix a ton of CO_2, you need 1.6 to 3.7 tons of rock. These rocks have to be mined and transported to coal plants (some industrial wastes and mining tailings can also be used, for example fuel ash from coal plants or de-inking ash from the paper industry, but their total amounts are too small to substantially reduce CO_2-emissions). The process also generates large amounts of waste materials (apart from the carbonized rocks themselves). For every ton of carbon dioxide stored in rock, you are left with 2.6 to 4.7 tons of disposable materials.

Carbon capture technology is expected to become more energy-efficient in the future. But that would hardly make the whole scheme more attractive. Storing carbon dioxide in underground reservoirs (the only realistic option) is risky, not unlike the storing of atomic waste.

CO_2 can escape. High concentrations of the gas are lethal to plants, animals and humans. Eventually the gas thins in the atmosphere but during escape concentrations it can build up fast, especially since CO_2 is denser than air. At concentrations above 2 percent in ambient air, carbon dioxide has a strong effect on respiration (the normal concentration in fresh air is 0.033 percent). At concentrations from 7 to 10 percent, it kills.

Small impurities in the gas make it lethal at even lower concentrations. Similar amounts in the soil kill vegetation and make groundwater unsuitable for drinking or irrigation. And of course, if CO_2 escapes from storage reservoirs, the whole operation of capturing, transporting and storing it is all for nothing. The result is a considerable rise in emissions, because of the energy penalty involved (the energy use of the whole process can go down, but it will never come close to zero).

We can draw an analogy that storing CO_2 in essence resembles storing a gas in a stone pot. When an underground reservoir is filled up, all injection wells are closed by a cement "cork". This cork has to keep the gas inside for thousands of years.

Now zoom out and you are looking at a planet with thousands of holes, each one sealed with a cork to keep inside a potentially deadly (and climate warming) gas. Reassuring, no? Yet, this is considered a high-tech solution. You also have a network of thousands of kilometers of pipelines, transporting the same gas and connecting power plants with these reservoirs.

How big is the chance that leaks or sudden bursts will cause damage? According to the IPCC, the risk is low, at least if all pipelines and storage sites are monitored closely. Leaks will occur and they also occur through cracks in natural CO_2-reservoirs (sometimes killing vegetation, animals and people), but if watched closely (by means of sensors and computers) they could be stopped in time. Earthquakes or accidental drilling operations in a former storage site could release large clouds of the gas in a short time, but that should be prevented by using carefully chosen locations and painstaking indexing of storage sites.

All these risks might be manageable in the short term, but storing CO_2 is — just like storing atomic waste — a very irresponsible thing to do in respect to future generations. Will people in 2178 still know where CO_2 was stored? Will the corks hold until that date? Risk analysis does not seem to look too far ahead. Carbon capture will also bury a part of our oxygen supply, by the way. After all, we would not only store the C but also the O_2. Oxygen is abundant in the atmosphere compared to carbon, so this might not be a big problem, but again, nobody seems to have investigated this yet.

Carbon storage in the oceans was radically disapproved of by the IPCC because of the even larger risks, yet the idea returned last week and seems to be gaining support even with green thinkers.

Real Solutions, Please

Why introduce yet another expensive, energy-intensive and risky technology if there are so many other and better ways to solve the energy crisis? If we choose to build a whole new infrastructure of pipelines comparable to that of the existing oil and gas industry, why not build something like an extensive underground tubular freight network instead? This would be a real solution, which would considerably lower transport energy use and CO_2-emissions.

Why not channel the huge amounts of money needed for the development of CCS to countries with tropical rainforests, so that they have a very good reason to protect them fiercely? Stopping deforestation, especially in tropical forests, would contribute more to the fight against global warming than carbon capture technology could ever do. Tropical forests store enormous amounts of carbon and they are not prone to natural forest fires.

Why not put into force a regulation that prohibits the construction of any more power plants that burn non-renewable energy sources? There is already an enormous energy capacity in the world, why don't we choose to do it with the energy plants that we have? This would at last make energy efficiency useful (because progress in energy efficiency is now always negated by new and more energy hungry products and services). Still want more energy? Build a solar plant or plant a windmill.

These are just 3 ideas that would be effective without the need to adapt our lifestyle (which is, of course, also the attraction of "clean coal"). They would not solve everything, but at least they would be very welcome steps into the right direction, towards a solution.

All high-tech carbon storage strategies described in this article are no solutions, which are just attempts to limit the problem. Let's hope that the next appeal of the International Energy Association and of the Science Academies of the world (an awful lot of brains there) will contain a trace of innovative thinking.

Additional Words and Phrases

biodiversity: 生物多样性
acid fallout: 酸性沉降物
acid fog: 酸雾
acid rain: 酸雨

adverse climate change：不利的气候变化
arctic ozone hole：北极臭氧层空洞
atmospheric acidity：大气酸度
atmospheric circulation pattern：大气环流模式
atmospheric ozone：大气臭氧
atmospheric transformation product：使大气变性的产物
biochemical oxygen demand (BOD)：生物氧化量
chemical cleaning of coal：煤的化学洁净
chemical substitute：化学替代物质
circulation pattern of the atmosphere：大气环流模式
clean use of coal：煤的洁净用法
cleaning solvent：洗涤液，洗涤溶剂
climate sensitivity：气候敏感性
climatic upheaval：气候剧烈变化
climatic variability：气候可变性
collection of household refuse：家庭垃圾收集
combustion emissions：燃烧废气
consumption area (of ozone depleting substances)：消耗臭氧物质的消费领域
desulphurization：脱硫作用
disaster management：灾害管理
disaster-prone area：易受灾地区
discharge into the environment：排入环境中
disposal capabilities：废物处理能力
dissemination in the air：在空气中散布
drought early warning：干旱预警
dumping (of hazardous wastes)：(危险物的)倾弃
ecological disaster area：生态灾区
ecological disruption：生态失调
ecological pyramid：生态金字塔
ecological rehabilitation：生态复原
ecological security：生态安全
ecological suitability：生态适应性
ecologically safe product：生态无害产品
ecologically sound characteristics (of a product)：(一种产品的)生态无害特性

ecologically sound product：生态无害产品
ecologically sound technology：生态无害技术
ecosystem rehabilitation：生态系复原
ecosystem approach：生态系观点
ecotechnology：生态技术；生态无害技术
ecotone：群落交错区；群落过渡区
ecotoxicology：生态毒理学
effluent charge：排污费；排污税
effluent discharge：排放废物；排出废液
effluent fee：排污费；排污税
effluent standard：排放标准
energy-intensive：能源密集的
environmental constraints：环境制约；环境约束
environmental degradation：环境退化
environmental despoliation：环境掠夺
environmental deterioration：环境恶化
environmental disaster：环境灾难
environmental disorder：环境失调；环境混乱
environmental dispute：环境争论；环境纠纷
environmental distribution：环境分配
environmental economics：环境经济学
environmental education：环境教育
environmental efficiency：环境效率
environmental emergency：环境紧急事故
exhaust gas recirculation (ERG)：废气回流；废气还流；废气再循环装置
fuel economy penalty：燃料经济性的缺陷
gas-borne particles：悬浮气体中的微粒
greenhouse gas (GHG)：温室气体
high-sulphur fuel：高硫燃料
high-sulphur residues：高硫残留物
incineration plant：焚烧车间
industrial effluent treatment plant：工业废水处理厂
land clearance：清理土地；清除地物；开荒
land disposal of sewage sludge：下水污泥的地面处理

land disposal site：地面处理场
land erosion control：土地侵蚀控制；土地侵蚀治理
landfill（site）：填地；垃圾填埋地；填埋
landfill dumping：倾弃垃圾填埋
landfill treatment：垃圾填埋处理
powder waste：粉末状废物
powdery substance：粉末状物质
resource recovery：资源回收
resource recovery plant：资源回收厂；回收车间
respirable suspended particulates（RSPs）：可因呼吸进入人体的悬浮微粒
respirable-sized particulates：颗粒大小；可因呼吸进入人体的微粒
sediment：沉降物；沉积物；沉淀物
sediment core：沉积岩心
sedimentation：沉降作用；沉积作用；沉淀作用
sedimentation basin：沉淀池；沉积盆地
sedimentation rate：沉降速度
sedimentation tank：沉降槽；沉降池
activated carbon：活性煤
advance borehole：超前钻孔
bench stoping：阶梯式开采
hauling capacity：运输能力
hoisting equipment：提升设备
hollow concrete：空心混凝土
horizontal slicing：水平分层开采法
tailings disposal plant：尾矿处理装置
thermomagnetic preparation：热磁选煤法
tunneling equipment：隧道掘进设备

Unit 8　Clean Coal Technology

Foreword

Coal is the dirtiest of all fossil fuels. When burned, it produces emissions that contribute to global warming, create acid rain and pollute water. With all of the hoopla surrounding nuclear energy, hydropower and biofuels, you might be forgiven for thinking that grimy coal is finally on its way out. But coal is no sooty remnant of the Industrial Revolution — it generates half of the electricity in the United States and will likely continue to do so as long as it's cheap and plentiful. Clean coal technology seeks to reduce harsh environmental effects by using multiple technologies to clean coal and contain its emissions.

Section One

Preview

Text A offers you an overview of clean coal technologies available now. The text begins with an analysis of the pollutants produced by coal mining, such as SO_2 and CO_2. Then by a brief induction to the clean coal technologies that are being developed or utilized in the world, we can safely conclude that it is still promising for us to reduce the gas emissions, thus creating a better environment. Finally, by citing some successful projects, the text provides certain guidance for any company or government on how to execute their environment-friendly strategies.

Text A

Clean Coal Technology

"Clean" coal technology is a collection of technologies being developed to

mitigate the environmental impact of coal energy generation. When coal is used as a fuel source, the **gaseous** emissions generated by the **thermal decomposition** of the coal include sulphur dioxide (SO_2), nitrogen oxides (NOx), carbon dioxide (CO_2), mercury, and other chemical byproducts that vary depending on the type of the coal being used. These emissions have been established to have a negative impact on the environment, contributing to acid rain and other pollutants. As a result, clean coal technologies are being developed to remove or reduce pollutant emissions to the atmosphere. Some of the techniques that would be used to accomplish this include chemically washing minerals and **impurities** from the coal, **gasification**, improved technology for treating flue gases to remove pollutants to increasingly **stringent** levels and at higher efficiency, carbon capture and storage technologies to capture the carbon dioxide from the **flue gas** and dewatering lower rank coals (brown coals) to improve the **calorific** value, and thus the efficiency of the conversion into electricity. Figures from the United States Environmental Protection Agency show that these technologies have made today's coal-based generating fleet 77 percent cleaner on the basis of regulated emissions per unit of energy produced.

Clean coal technology usually addresses atmospheric problems resulting from burning coal. Historically, the primary focus was on SO_2 and NOx, the most important gases in **causation** of acid rain, and **particulates** which cause visible air pollution and **deleterious** effects on human health. More recent focus has been on carbon dioxide (due to its impact on global warming) and concern over **toxic** species such as mercury. Concerns exist regarding the economic **viability** of these technologies and the timeframe of delivery, potentially high hidden economic costs in terms of social and environmental damage, and the costs and viability of disposing of removed carbon and other toxic matter.

Several different technological methods are available for the purpose of carbon capture as demanded by the clean coal concept:

Pre-combustion capture. This involves gasification of a **feedstock** (such as coal) to form **synthesis** gas, which may be shifted to produce a H_2 and CO_2-rich gas mixture, from which the CO_2 can be efficiently captured and separated, transported, and ultimately **sequestered**. This technology is usually associated with **Integrated** Gasification Combined Cycle process **configurations**.

Post-combustion capture. This refers to capture of CO_2 from exhaust gases of combustion processes, typically using **sorbents**, **solvents**, or **membrane** separations to remove CO_2 from the **bulk** gases.

Oxy-fuel combustion. Fossil fuels such as coal are burned in a mixture of **recirculated** flue gas and oxygen, rather than in air, which largely **eliminates** nitrogen from the flue gas enabling efficient, low-cost CO_2 capture.

The Kemper County IGCC Project, a 582 MW coal gasification-based power plant, will use pre-combustion capture of CO_2 to capture 65% of the CO_2 the plant produces, which will be utilized / geologically sequestered in enhanced oil recovery operations.

The Saskatchewan Government's Boundary Dam Integrated Carbon Capture and Sequestration Demonstration Project will use post-combustion, **amine**-based **scrubber** technology to capture 90% of the CO_2 emitted by the power plant; this CO_2 will be **pipelined** to and utilized for enhanced oil recovery in the Weyburn oil fields.

An oxy-fuel CCS power plant operation processes the exhaust gases so as to separate the CO_2 so that it may be stored or sequestered.

An early example of a coal-based plant using (oxy-fuel) carbon-capture technology is Swedish company Vattenfall's Schwarze Pumpe power station located in Spremberg, Germany, built by German firm Siemens, which went on-line in September 2008. The facility captures CO_2 and acid rain producing pollutants, separates them, and compresses the CO_2 into a liquid. Plans are to inject the CO_2 into **depleted** natural gas fields or other geological formations. Vattenfall **opines** that this technology is considered not to be a final solution for CO_2 reduction in the atmosphere, but provides an achievable solution in the near term while more desirable **alternative** solutions to power generation can be made economically practical.

Other examples of oxy-combustion carbon capture are in progress. In Biloela, Queensland, Australia, Callide Power Station has **retrofitted** a 30-MWth existing PC-fired power plant to operate in oxy-fuel mode; in Ciuden, Spain, Endesa has a newly built 30-MWth oxy-fuel plant using circulating **fluidized** bed combustion (CFBC) technology. Babcock-ThermoEnergy's Zero Emission Boiler System (ZEBS) is oxy-combustion-based; this system **features** near 100% carbon-capture and according to company information there is virtually no air-emissions.

Other carbon capture and storage technologies include those that **dewater** low-rank coals. Low-rank coals often contain a higher level of **moisture** content which contains a lower energy content per tonne. This causes a reduced burning efficiency and an increased emissions output. Reduction of moisture from the coal prior to combustion can reduce emissions by up to 50 percent.

The UK government is working towards a clean energy future and supports clean coal projects across the country. In August 2010, UK-based company B9 Coal announced a clean coal project with 90% carbon capture to be put forward to DECC. This would help the UK raise its profile amongst green leaders across the world. This proposed project gasifies coal underground and processes it to create pure streams of hydrogen and carbon dioxide. The hydrogen is then used as an emissions-free fuel to run an **alkaline** fuel cell whilst the carbon dioxide is captured. This UK project could provide a world-leading **template** for clean coal with CCS globally.

According to United Nations Intergovernmental Panel on Climate Change, the burning of coal, a fossil fuel, is a major contributor to global warming. As 25.5% of the world's electrical generation in 2004 was from coal-fired generation, reaching the carbon dioxide reduction targets of the Kyoto Protocol will require **modifications** to the way coal is utilized.

Coal, which is primarily used for the generation of electricity, is the second largest domestic contributor to carbon dioxide emissions in the USA. The public has become more concerned about global warming which has led to new legislation. The coal industry has responded by running advertising **touting** clean coal in an effort to counter negative **perceptions** and claiming more than $50 billion towards the development and deployment of "traditional" clean coal technologies over the past 30 years; and promising $500 million towards carbon capture and storage research and development.

Vocabulary

hoopla / ˈhuːplɑː / **n.** great fuss or excitement 喧闹

biofuel / ˈbaɪəʊˈfjuːəl / **n.** a gas, liquid, or solid from natural sources such as plants that is used as a fuel 生物燃料

Unit 8　Clean Coal Technology

grimy / ˈgraɪmɪ / *adj.* thickly covered with ingrained dirt 肮脏的
sooty / ˈsʊtɪ / *adj.* of the blackest black 乌黑的
remnant / ˈremnənt / *n.* a small part or portion that remains after the main part no longer exists 剩余（物）
emission / ɪˈmɪʃ(ə)n / *n.* letting something out 发射；排放
utilize / ˈjuːtəlaɪz / *v.* to make use of something 利用
mitigate / ˈmɪtɪgeɪt / *v.* to lessen 缓和；缓解
gaseous / ˈgæsɪəs；ˈgeɪsɪəs / *adj.* containing gas 气态的；气体的
thermal / ˈθɜːm(ə)l / *adj.* involving heat 热的；热量的；保热的
decomposition / ˌdiːkɒmpəˈzɪʃn / *n.* breaking down into pieces 分解；腐烂
impurity / ɪmˈpjʊərɪtɪ / *n.* lack of purity 杂质；不纯
gasification / ˌgæsəfəˈkeʃən / *n.* the process of changing into gas 气化
stringent / ˈstrɪn(d)ʒ(ə)nt / *adj.* rigorous or severe 迫切的
flue gas / fluː / *n.* the smoke in the uptake of a boiler fire that consists mainly of carbon dioxide, carbon monoxide, and nitrogen 废气
calorific / kæləˈrɪfɪk / *adj.* of heat; full of calories 发热的；生热的
causation / kɔːˈzeɪʃ(ə)n / *n.* cause or reason 原因
particulate / pɑːˈtɪkjʊlət; -eɪt; pə- / *adj.* of particles 微粒的
deleterious / ˌdelɪˈtɪərɪəs / *adj.* harmful 有毒的；有害的
toxic / ˈtɒksɪk / *adj.* poisonous 有毒的
viability / ˌvaɪəˈbɪlətɪ / *n.* capable of being done in a practical and useful way 可行性
feedstock / ˈfiːdstɒk / *n.* raw material 原料
synthesis / ˈsɪnθɪsɪs / *n.* result of combining 合成
sequester / sɪˈkwestə / *v.* to isolate 使隔绝
integrate / ˈɪntɪgreɪt / *v.* to make into a whole 整合
configuration / kənˌfɪgəˈreɪʃ(ə)n; -gjʊ- / *n.* arrangement of parts 配置
sorbent / ˈsɔːb(ə)nt / *n.* a material used to absorb or adsorb liquids or gases 吸附剂
solvent / ˈsɒlv(ə)nt / *n.* a substance, especially a liquid, that can dissolve another substance 溶剂
membrane / ˈmembreɪn / *n.* thin layer of something 膜
bulk / bʌlk / *n.* majority 大部分
oxy-fuel *n.* technology that burns pure oxygen with gaseous fuel 氧燃料

recirculate / riːˈsɜːkjʊleɪt / *v.* to circulate again 再流通
eliminate / ɪˈlɪmɪneɪt / *v.* to get rid of 消除；排除
amine / əˈmiːn / *n.* any organic derivative of ammonia formed by the replacement of hydrogen with one or more alkyl groups 胺
scrubber / ˈskrʌbə / *n.* a brush or other object that you use for cleaning things, for example pans（刷洗用的）刷子；洗涤器
pipeline / ˈpaɪplaɪn / *v.* transmit or transport in a pipeline 用管道运输
deplete / dɪˈpliːt / *v.* to use up 耗尽，用光
opine / ə(ʊ)ˈpaɪn / *v.* to state opinion 以为
alternative / ɔːlˈtɜːnətɪv; ɒl- / *adj.* mutually exclusive 供选择的
retrofit / ˈretrəʊfɪt / *v.* to modify 改进
fluidize / ˈfluːɪdaɪz / *v.* to cause to be fluid 使流体化
feature / ˈfiːtʃə / *v.* contain something as important element 以……为特色
dewater / diːˈwɔːtə / *v.* to remove water from 脱水
moisture / ˈmɔɪstʃə / *n.* very small drops of water that are present in the air, on a surface or in a substance 湿气
alkaline / ˈælkəlaɪn / *adj.* having the nature of an alkali 碱性的
template / ˈtempleɪt; -plɪt / *n.* pattern 模板
modification / ˌmɒdɪfɪˈkeɪʃ(ə)n / *n.* act of modifying 修正；修改
tout / taʊt / *v.* to try to sell something 兜售；招揽
perception / pəˈsepʃ(ə)n / *n.* view or opinion 看法

Exercises

Ⅰ. **Comprehension of the Text**

Directions: Please answer the following questions according to the text.

1. What is clean coal technology?
2. What are the main pollutants released from coal burning?
3. What has caused the delay in implementing clean coal technologies? And why?
4. How is CO_2 trapped and processed in the technology of pre-combustion capture?
5. Is it possible for carbon-capture (oxy-fuel) technology to be widely spread? If not, why?

6. What are the other carbon capture and storage technologies mentioned in the passage?
7. What has the UK government done to achieve the goal of a clean energy future?
8. What has the US coal industry done to neutralize the public negative perceptions about coal?

II. Group Discussion

Directions: Discuss the following question with your partners, using as much as text information as possible in your discussion.

1. Do you think it's possible for the coal industry in China to adopt these clean coal technologies in the foreseeable future?
2. What do you think is most important in spreading clean coal technologies around the world?
3. If you were the CEO of a coal company, what would you do to embrace clean coal technologies while reducing the costs?

III. Word Bank

Directions: Fill the blanks with the words on the right side of the Text. For each word, you can use only once.

Burning coal can release a lot of 1. _____ causing visible air pollution and 2. _____ effects on human health. In recent years, in an effort to 3. _____ the impact of coal burning on the environment, people around the world are trying all possible means to develop clean coal technologies to reduce CO_2 4. _____. These new technologies have proved to be quite successful and some of them can even remove pollutants to 5. _____ levels. For example, the Kemper County IGCC Project will use pre-combustion capture of CO_2 to capture 65% of the CO_2 the plant produces, which in turn largely 6. _____ nitrogen from the flue gas. Other technologies include those that

a) viability
b) dewater
c) perceptions
d) deleterious
e) particulates
f) template
g) remnant
h) alternative
i) emissions
j) mitigate
k) pine
l) eliminates
m) utilize
n) stringent
o) feature

7. _____ low-rank coal. Some governments have also stepped in to improve the environment. The UK government has initiated a lot of programs for a clean energy future, hoping to provide a world-leading 8. _____. The US coal industry is trying to counter negative 9. _____ about coal's destructive impact on the environment by running advertising. But due to the economic 10. _____ of these technologies, some people are worried that if these new technologies will be welcomed by the industry.

IV. Translation Practice

Directions: Translate the following sentences from English to Chinese.

1. These emissions have been established to have a negative impact on the environment, contributing to acid rain and other pollutants. As a result, clean coal technologies are being developed to remove or reduce pollutant emissions to the atmosphere.
2. Figures from the United States Environmental Protection Agency show that these technologies have made today's coal-based generating fleet 77 percent cleaner on the basis of regulated emissions per unit of energy produced.
3. Historically, the primary focus was on SO_2 and NOx, the most important gases in causation of acid rain, and particulates which cause visible air pollution and deleterious effects on human health.
4. Fossil fuels such as coal are burned in a mixture of recirculated flue gas and oxygen, rather than in air, which largely eliminates nitrogen from the flue gas enabling efficient, low-cost CO_2 capture.
5. Vattenfall opines that this technology is considered not to be a final solution for CO_2 reduction in the atmosphere, but provides an achievable solution in the near term while more desirable alternative solutions to power generation can be made economically practical.
6. As 25.5% of the world's electrical generation in 2004 was from coal-fired generation, reaching the carbon dioxide reduction targets of the Kyoto Protocol will require modifications to the way coal is utilized.

V. Writing Practice

Directions: Write a 3-paragraph passage of about 120 words with the title "Clean Coal Technology: an Environmental Necessity" based on the outline below.

1. Destructive exploitation of coal has led to severe environmental impact in the past decades.
2. Technological advances in coal mining can spur wide changes whether in our energy policies or in improvement of efficiency.
3. What might be expected in the future?

Section Two

Preview

CO_2 is a major contributor to the worsening global environment. It is responsible for many environmental catastrophes. Burning coal, the dominant source for our energy now or in the near future, produces billion tons of CO_2 each year. Then how to capture and separate CO_2 in the process of burning coal is becoming not only a technological priority but also a necessity for the world's sustainable development.

Text B

Capture and Separation of CO_2

Coal is an extremely important fuel and will remain so. Some 23% of primary energy needs are met by coal and 39% of electricity is generated from coal. About 70% of world steel production depends on coal feedstock. Coal is the world's most **abundant** and widely distributed fossil fuel source. The International Energy Agency (IEA) expects a 43% increase in its use from 2000 to 2020.

However, burning coal produces almost 14 billion tons of carbon dioxide each year which is released to the atmosphere, most of this being from power generation.

Development of new "clean coal" technologies is addressing this problem so that the world's enormous resources of coal can be utilized for future generations without contributing to global warming. Much of the challenge is in commercializing the technology so that coal use remains economically competitive despite the cost of achieving low, and eventually "near-zero", emissions.

As many coal-fired power stations approach retirement, their replacement gives much scope for "cleaner" electricity. Alongside nuclear power and harnessing renewable energy sources, one hope for this is via "clean coal" technologies, such as carbon capture and sequestration (CCS). However in its 2014 Energy Technology Perspectives the IEA notes that "CCS is advancing slowly, due to high costs and lack of political and financial commitment. Few major developments were seen in 2013, and policies necessary to **facilitate** the transition from demonstration to deployment are still largely missing." For its low-carbon 2DS **scenario**, "the rate of capture and storage must increase by two orders of **magnitude**" by 2025.

A number of means exist to capture carbon dioxide from gas streams, but they have not yet been **optimized** for the scale required in coal-burning power plants. The focus in the past has often been on obtaining pure CO_2 for industrial purposes rather than reducing CO_2 levels in power plant emissions.

Where there is carbon dioxide mixed with methane from natural gas wells, its separation is well proven. Several processes are used, including hot **potassium carbonate** which is energy-intensive and requires a large plant, a **monoethanolamine** process which yields high-purity carbon dioxide, amine scrubbing, and membrane processes.

Development of CCS for coal combustion has lost **momentum** in the last few years, partly due to uncertainty regarding carbon emission prices.

In mid-2010 the IEA published a report that says CCS was challenging, and quoting $26 billion committed in the previous two years to CCS projects. There were 80 large-scale integrated CCS projects under way, 5 of them operating. It said that "notable efforts" were being made and "increased action", but "rapid progress is now required" if CCS is to be deployed by 2020.

Post-combustion Capture

Capture of carbon dioxide from flue gas streams following combustion in air is much more difficult and expensive than from natural gas streams, as the carbon dioxide concentration is only about 14% at best, with nitrogen most of the rest, and the flue gas is hot. The main process treats carbon dioxide like any other pollutant, and as flue gases are passed through an amine solution the CO_2 is absorbed. It can later be released by heating the solution. This amine scrubbing process is also used for taking CO_2 out of natural gas. There is a significant energy cost involved. For new power plants this is quoted as 20-25% of plant output, due both to reduced plant efficiency and the energy requirements of the actual process.

No commercial-scale power plants are operating with this process yet. At the new 1300 MWe (兆瓦电) Mountaineer power plant in West Virginia, less than 2% of the plant's off-gas is being treated for CO_2 recovery, using chilled amine technology. This has been successful. Subject to federal grants, there are plans to capture and sequester 20% of the plant's CO_2, some 1.8 million tons CO_2 per year.

Oxyfuel Combustion

Where coal is burned in oxygen rather than air, it means that the flue gas is mostly CO_2 and hence it can more readily be captured by amine scrubbing — at about half the cost of capture from conventional plants. A number of oxyfuel systems are **operational** in the USA and elsewhere, and the FutureGen (未来发电项目) 2 project involves oxy-combustion. Such a plant has an air separation unit, a boiler island, and a compression and purification unit for final flue gas.

The Integrated Gasification Combined Cycle (IGCC) plant is a means of using coal and steam to produce hydrogen and carbon monoxide (CO) from the coal and these are then burned in a gas turbine with **secondary** steam turbine (ie combined cycle) to produce electricity. If the IGCC gasifier is fed with oxygen rather than air, the flue gas contains highly-concentrated CO_2 which can readily be captured post-combustion as above.

In China, the first phase of Huaneng Group's \$1.5 billion GreenGen (绿色发电项目) project is a 250 MWe oxyfuel IGCC power plant burning hydrogen and carbon monoxide which is due to **commence** operation by mid 2012. A second phase involves a pilot plant to produce electricity from hydrogen. Phase 3 will be a 400 MWe commercial plant with CCS.

Pre-combustion Capture

Further development of the IGCC process will add a shift reactor to **oxidase** the CO with water so that the gas stream is basically just hydrogen and carbon dioxide, with some nitrogen. The CO_2 with some H_2S and Hg impurities are separated before combustion (with about 85% CO_2 recovery) and the hydrogen alone becomes the fuel for electricity generation (or other uses) while the concentrated **pressurized** carbon dioxide is readily disposed of. (The H_2S is oxidized to water and sulfur, which is **saleable**.) No commercial-scale power plants are operating with this process yet.

Currently IGCC plants typically have a 45% thermal efficiency.

Capture of carbon dioxide from coal gasification is already achieved at low **marginal** cost in some plants. One (**albeit** where the high capital cost has been largely written off) is the Great Plains Synfuels Plant in North Dakota, where 6 million tonnes of lignite is gasified each year to produce clean synthetic natural gas.

Oxy-fuel technology has potential for retrofit to existing **pulverized** coal plants, which are the backbone of electricity generation in many countries.

In China, the major utility China Datang Corp is teaming with Alstom to build two demonstration CCS projects. A 350 MWe coal-fired plant at Daqing, Heilongjiang province, will be equipped with Alstom's oxy-firing technology, and a 1000 MWe coal-fired plant at Dongying, Shandong province, will use an Alstom's post-combustion capture technology, either chilled **ammonia** or advanced amines. The two projects are expected to be operational in 2015 and each capture over one million tons of CO_2 per year, which would be about 40% of output from Daqing and 15% from Dongying, though Alstom says that the actual levels of capture and storage have not yet been defined and will be in the scope of the first **feasibility** studies of the respective projects. **Adjacent** oilfields will be used for sequestration, enabling enhanced oil recovery.

Vocabulary

catastrophe / kəˈtæstrəfɪ / *n.* disaster 大灾难
dominant / ˈdɒmɪnənt / *adj.* in control 占优势的
priority / praɪˈɒrɪtɪ / *n.* right of precedence 优先

Unit 8 Clean Coal Technology

sustainable / sə'steɪnəb(ə)l / *adj.* able to be maintained 可持续的
abundant / ə'bʌnd(ə)nt / *adj.* plentiful 丰富的
facilitate / fə'sɪlɪteɪt / *v.* to simplify process 使容易
scenario / sɪ'nɑːrɪəʊ / *n.* possible situation 方案
magnitude / 'mægnɪtjuːd / *n.* greatness of size 量级
optimize / 'ɒptɪmaɪz / *v.* to enhance effectiveness of something 使最优化
potassium carbonate / pə'tæsɪəm / / 'kɑːbəneɪt / *n.* a white salt 磷酸钾
monoethanolamine / ˌmɒnəʊˌeθənə'læmiː n / *n.* 单乙醇胺
momentum / mə'mentəm / *n.* capacity for progressive development 势头；动力
operational / ˌɒpə'reɪʃ(ə)n(ə)l / *adj.* of operating 操作的；运作的
secondary / 'sek(ə)nd(ə)rɪ / *adj.* not primary or major 次级的
commence / kə'mens / *v.* to begin 开始
oxidase / 'ɒksɪdeɪz / *n.* an enzyme that catalyzes oxidation 氧化酶
pressurize / 'preʃəraɪz / *v.* to increase air pressure in container 密封
saleable / 'seɪləb(ə)l / *adj.* available to be sold 畅销的；可供出售的
marginal / 'mɑːdʒɪn(ə)l / *adj.* small in scale 边缘的；临界的
albeit / ɔːl'biːɪt / *conj.* even though 虽然
pulverize / 'pʌlvəraɪz / *v.* to crush something to powder 粉碎
ammonia / ə'məʊnɪə / *n.* a gas with a strong smell; a clear liquid containing ammonia, used as a cleaning substance 氨
feasibility / ˌfiːzɪ'bɪlɪtɪ / *n.* practicality 可行性
adjacent / ə'dʒeɪs(ə)nt / *adj.* neighboring 邻近的
amine scrubbing 氨净化

Exercises

Directions: Choose the right answer to each question according to the text.

1. According to the passage, the main barrier to the wide-spread promotion of clean coal technology is _____.
   A. increasing environmental effect of coal
   B. commercialization of these technologies
   C. opposition from the governments
   D. severe competition from other sources of energy
2. The 2014 Energy Technology Perspectives released by IEA suggested that

_____ .

A. CCS is our focus in technological development in the future.

B. Many coal-fired power stations are reluctantly to adopt new technologies because they lack resources.

C. There has been much progress made in applying these CCS technologies.

D. There is urgent need to CCS development and promotion

3. A lot of reasons account for the slow development of CCS EXCEPT _____ .

A. These technologies are only tested or proved successful on a small scale

B. There are no polices to guide the power stations in adopting these technologies

C. These technologies might cost a lot of money

D. Some technologies may consume a lot of energy

4. Compared with carbon dioxide capture of natural gas, that of flue gas is _____ .

A. more efficient

B. quite different

C. more energy-consuming

D. more complex

5. Huaneng Group's GreenGen project is mentioned to prove that _____ .

A. China has made much progress in its commitment to reducing CO_2 emissions

B. CCS technology in oxyfuel combustion is commercially possible.

C. in comparison, some companies in China have done better job in adopting new technologies than in other countries

D. though the CO_2 capture in oxyfuel combustion is much more difficult than that in post-combustion, still it has entered commercial phase

6. Why are no commercial-scale power plants operating pre-combustion capture?

A. There are no such technologies available.

B. It can only capture a small amount of CO_2.

C. It is so costly that no commercial-scale power plants want to adopt it.

D. There are so many impurities.

7. From the passage, we can infer that some power plants try to experiment with these CCS technologies because _____.
   A. They are environmentally-conscious.
   B. They are willing to sacrifice their profits for the human welfare.
   C. Some organizations, such as governments, subsidize them.
   D. They realize that these technologies have great potential for future growth.
8. What does this passage say about the CCS projects in China Datang Corp?
   A. They are going to set a good example for the whole world to follow.
   B. They have proved to be quite successful.
   C. Whether these projects will be widely adopted is still uncertain, but hopefully so.
   D. They will not work out because of their economic cost.

Section Three

Extended Reading

Does "Clean Coal" Technology Have a Future?

The mix of energy sources used to produce electricity is changing — slowly. Coal is still king and is expected to retain that title for decades, giving ground only gradually to renewable fuels, natural gas and nuclear power.

Coal will account for 39% of global net electricity generation next year and 36% in 2040, according to projections by the U. S. Energy Information Administration.

Many people would like to see that number drop more dramatically. With concerns mounting about the effect of greenhouse gases on the global climate, pressure is growing for utilities to reduce carbon-dioxide emissions in their power production.

The power industry has responded in part by increasing its use of renewable energy sources. But it also continues to pursue another idea to help address environmental concerns: clean up coal-fired power plants. Technology that achieves that by capturing most of the carbon dioxide in a plant's emissions and then liquefying it for underground storage or for commercial use is just starting to be implemented.

Proponents of renewable fuels want utilities to focus instead on investing much more heavily in wind and solar power. The many billions of dollars it would take to implement clean-coal technology on a global scale won't do enough to lessen coal's environmental impact, they argue. That money, they say, should be going toward speeding the arrival of renewable energy as the new king of power generation.

Howard J. Herzog, senior research engineer at the MIT Energy Initiative at the Massachusetts Institute of Technology, says clean coal has an important role to play in the future of power generation. Richard Heinberg, a senior fellow-in-residence at the Post Carbon Institute, argues that investment should be directed instead to renewable energy sources.

Yes: Innovative Technology Will Rise to the Challenge

People have questioned the idea of clean coal for decades. It started with doubts about cleaning up particulate matter in power-plant emissions, then sulfur dioxide and nitrogen oxides, and it continues with skepticism about eliminating heavy metals like mercury.

Despite the naysayers, technology and innovation have risen to the challenge by providing effective and affordable solutions. Now, the challenge is to reduce coal's carbon-dioxide emissions. If history is any guide, innovative technology will once again provide the solutions.

The key technology needed to drastically reduce CO_2 emissions from coal-fired power plants is carbon capture and storage, or CCS. All of the components of this system are in commercial operation today. At this point, they are employed mostly to enhance oil recovery. There are about 4,000 miles of pipeline in the U.S. transporting tens of millions of tons of compressed CO_2 annually, mostly from natural wells, for injection into geologic formations to help extract oil. Numerous demonstration projects have shown that captured CO_2 also can be safely and effectively stored in deep geologic formations, as most of it will be.

Last month brought an important milestone: SaskPower's Boundary Dam power plant in Canada officially opened as the world's first commercial-scale coal power plant with CCS. About 90% of the plant's CO_2 is captured and piped about 40 miles for injection into oil fields. Next year, the Mississippi Power unit of Southern Co. will start operating a new clean-coal plant, and construction has just started on a clean-coal power plant in Texas. Other projects are being planned, most prominently in the U.S., U.K. and China.

Clean coal will become more common because climate policy will demand cleaner power. For instance, an emissions restriction on coal-fired power plants in Canada was a major driver for the Boundary Dam project. There will be added costs to power providers. But clean coal won't be so expensive that it can't compete with renewable or nuclear resources. All three will find significant markets. Yes, clean coal will require massive infrastructure investments on a global scale — but so will a major expansion of renewable-energy projects. For the electricity price of the Cape Wind project in Massachusetts, we could easily build a clean-coal plant with CCS.

A recently released assessment of the Intergovernmental Panel on Climate Change shows that clean-coal projects are projected to be competitive in a low-carbon world, and that excluding CCS from a mitigation-technology portfolio would more than double the cost of achieving climate-stabilization goals through 2100.

Selling captured carbon for enhanced oil recovery can help reduce the cost of CCS. And new technologies under development could allow carbon to be captured with dramatically lower expenditure of energy.

As for coal being a finite resource, that isn't a factor in the near term. We have centuries of coal supply. It is true that coal production in the U.S. has dipped recently, but this is due to competition from low-price natural gas. It has nothing to do with depletion.

Meanwhile, CCS isn't the only road to clean coal. State-of-the-art coal-fired power plants are being built with much higher efficiencies that result in a 20% reduction in CO_2 emissions per kilowatt-hour of electricity produced. And other technologies and regulations are mitigating the impacts of coal mining.

If you want to understand the energy industry, you must understand the power of innovation and technology. Instead of the oil shortages that some predicted, we have an oil glut thanks to technology. Many experts thought you couldn't profitably produce oil and gas from shale; technology proved otherwise. And it is because of technology that I'm optimistic about the future of clean coal.

No: The Economics Simply Don't Work — and Will Get Worse

For years, Americans have seen commercials touting "clean coal," while politicians on both sides of the aisle have extolled its promise. The technology to capture carbon emissions from coal-fired power plants has been tried and tested. Yet today almost none of the nation's coal-fueled plants are "clean."

Why the delay? The biggest problem for "clean coal" is that the economics don't work. Carbon capture and storage, or CCS, is extremely expensive. That gives the power industry little incentive to implement it in the absence of a substantial carbon tax.

Why would implementing CCS be so expensive? For starters, capturing and storing the carbon from coal combustion is estimated to consume 25% to 45% of the power produced, depending on the approach taken. That translates to not only higher prices for coal-generated electricity but also the need for more plants to serve the same customers. Other technologies designed to make carbon capture more efficient aren't commercial at this point, and their full costs are unknown.

And there's more. Capturing and burying just 38% of the carbon released from current U.S. coal combustion would entail pipelines, compressors and pumps on a scale equivalent to the size of the nation's oil industry. And while bolting CCS technology onto existing power plants is possible, it is inefficient. A new generation of plants would do the job much better — but that means replacing roughly 600 current-generation power plants.

Altogether, the Energy Department estimates that wholesale electricity prices with the initial generation of CCS technology would be 70% to 80% higher than current coal-based power.

The discussion of CCS technology in a recent assessment by the Intergovernmental Panel on Climate Change contains too many qualifiers to be interpreted as a declaration that clean coal will be competitive with renewable fuels.

Long term, the economics of coal are likely to get worse, with or without CCS. Coal is nonrenewable, finite in quantity and therefore subject to depletion. Rates of production from most regions of the U.S. are in decline.

And as depletion forces the mining of lower-quality resources, production prices will rise because of the need for more-sophisticated extraction technologies. Declining output is inevitable sooner or later.

Meanwhile, the price of electricity produced from solar and wind power is steadily dropping. The only thing that keeps coal-based electricity cheap today in relation to power from renewable sources is the industry's ability to shift the hidden costs — environmental and health damage — onto society. If, as climate regulations inevitably kick in, the coal power industry adopts CCS as a survival strategy, any lingering economic advantage over wind and even solar will disappear.

CCS also doesn't address the full range of coal's impact on society. It won't banish high rates of lung disease, because it doesn't eliminate all the pollutants from the combustion process or deal with the coal dust from mining and transport. It also doesn't address the environmental devastation of "mountaintop removal" mining.

This is not to say that "clean coal" has no future whatever. Coal plants with CCS will be built where captured carbon dioxide can be used to generate extra income — for example, by using it to stimulate old oil wells or make cement. But even a dramatic increase in such uses would put only a small fraction of carbon from coal to work.

A full transition of today's coal power industry to CCS is extremely unlikely unless the economics substantially change for some currently unforeseeable reason. And other technological advances, like more-efficient coal-fired plants, can only slow the growth of harmful emissions at best.

In all likelihood, the real future lies elsewhere — with distributed renewable energy.

Additional Words and Phrases

gaseous emissions：气态排放（物）
thermal decomposition：热分解
sulphur dioxide：二氧化硫
nitrogen oxides：氮氧化物
chemical byproducts：化工副产品

desulfurization process：脱硫工艺（过程）

gasification：气化

carbon capture and storage：碳捕获和储存

pre-combustion capture：燃烧前捕获

post-combustion capture：燃烧后捕获

oxy-fuel combustion：氧燃料燃烧技术

amine-based scrubber technology：基于氨的净化技术

carbon capture and sequestration：碳捕获与存储

landfill：垃圾场

biodegradable packaging：可降解包装

non-renewable：不可再生的

damage natural habitat：破坏自然栖息地

deforestation：森林消失

carbon dioxide：二氧化碳

acid rain：酸雨

greenhouse effect：温室效应

residue：废渣

industrial solid wastes：工业固体废物

white pollution：白色污染

organic pollutants：有机污染物

greenhouse effect：温室效应

Low-Carbon Economy (LCE)：低碳经济

Low-Carbon Life：低碳生活

Low-Carbon Tour：低碳旅游

Low-Carbon urbanization way：低碳城市化道路

the output of the carbon dioxide：二氧化碳排放

atmospheric concentrations of carbon dioxide：二氧化碳浓度

carbon emission reduction：碳减排

standard for carbon dioxide emission：碳排放标准

liability for carbon dioxide emission：碳排放责任

trans-frontier carbon dioxide pollution：越境碳污染

carbon monoxide：一氧化碳

greenhouse gas (GHS)：温室气体

greenhouse gas emission：温室气体排放

greenhouse effect：温室效应
biosphere：生物圈
ozone layer：臭氧层
ultraviolet ray：紫外线
infrared：红外线
ecosystem：生态系统
industrial fumes：工业烟尘
environmental accounting：环境核算
environmental auditing：环境审计
environmental health impact assessment：环境健康影响评价
environmental impact：环境影响
environmental impact assessment：环境影响评价
environmental impact statement：环境影响报告书
environmental indicators：环境指标
environmental policy：环境政策
environmental risk assessment：环境风险评估
sustainable development：可持续发展